BRITISH MUSEUM (NATURAL HISTORY)

A HANDBOOK ON EVOLUTION

by

GAVIN de BEER

Fourth Edition

Trustees of the British Museum (Natural History)
London : 1970
Price Seven Shillings (35p)

First Edition....... 1958
Second Edition........................1959
Third Edition 1964
Fourth Edition 1970

Printed and bound in England by
STAPLES PRINTERS LIMITED
at their Kettering, Northants, establishment

Preface

Sir Gavin de Beer, one of the world's foremost scholars on Darwin and an authority on evolution by natural selection, was Director of the British Museum (Natural History) from 1950 to 1960. It was during his tenure of office that he designed the display in the main hall of the Museum. The visitor could therefore have no better guide than this handbook. The handbook has, moreover, been written in such a way as to provide an introduction to evolution independently of the exhibits which it explains.

This fourth edition differs from the third in that the Geological Time-Scale (p. 123) has been revised.

G. F. CLARINGBULL
Director

Introduction

The series of exhibits illustrating Evolution was prepared to commemorate the Centenary of the first publication on the subject by Charles Darwin and Alfred Russel Wallace, and it remains on permanent exhibition to assist visitors to the Museum and students to acquire a grasp of the main principles involved in this epoch-making contribution to science.

This handbook is intended as a complement to the exhibits and is divided into three parts. In the first, a brief introduction is given to the theory of evolution by natural selection, accompanied by only such examples of observations and experiments as are necessary to provide a framework of the evidence on which it is based. This part was first published in abridged form in *Endeavour*, the editor of which has kindly agreed that it should be used here.

The second part is devoted to a brief description of the various Bays containing the exhibits. From the countless examples which might have been given, these exhibits have been selected so as to show a minimum of the evidence on which the general principles of the theory of evolution by natural selection rest.

The third part contains a brief account of six steps of major importance in evolution from the molecule to man.

Appended is a list of works suggested for general reading and references to specialized literature for the use of students.

Cross-references are given between Parts I, II, and III to facilitate the finding of exhibits illustrating particular points, and of the principles underlying particular exhibits. Names of persons in brackets after the descriptions of certain exhibits are those of scientists on whose recent work the exhibits are based, and I should particularly like to thank Dr. K. P. Oakley and Dr. W. T. Stearn for their help in connexion with the evolution of man and the history of cultivated cereals respectively. In the preparation of the exhibits on the general theory of evolution, the plans have been carried out, as regards design and most of the execution, by Mr. Erik Thorn. The remainder of the exhibits were designed and carried out by the staff of the Museum.

GAVIN DE BEER

Contents

Part I

EVOLUTION BY NATURAL SELECTION

Part II

THE EXHIBITS ILLUSTRATING EVOLUTION

East Side: First Bay
Right-hand Case

VARIATION

Facing Case

GENETIC VARIATION

East Side: Third Bay
Right-hand Case

ARTIFICIAL SELECTION

Facing Case

NATURAL SELECTION

Left-hand Case

SEXUAL SELECTION

East Side: Fourth Bay
Right-hand Case

CONTINUITY OF LIFE

Facing Case

LINES OF EVOLUTION

Left-hand Case

THE GEOLOGICAL SUCCESSION

East Side: Fifth Bay
Right-hand Case

CONSTANCY OF PLAN

Facing Case

GEOGRAPHICAL VARIATION

Left-hand Case

ISOLATION

West Side: First Bay
Left-hand Case

CLASSIFICATION

Right-hand Case

THE BONY FISHES

West Side: Second Bay
Left-hand Case

THE EVOLUTION OF MAN

Right-hand Case

THE EMERGENCE OF MODERN MAN

West Side: Third Bay
Left-hand Case

Facing Case

Right-hand Case

West Side: Fourth Bay
Left-hand Case

Right-hand Case

Part III

THE SIX MILESTONES OF EVOLUTION

West Side: Fifth Bay
Left-hand Case

List of Illustrations

EVOLUTION BY NATURAL SELECTION

From Special Creation to Transformism

Only one hundred years have gone by since the concept of evolution was brought to the attention of thinking men in a manner which has compelled its acceptance. The demonstration that the members of the plant and animal kingdoms are as they are because they have become what they are, and that change, not immutability, is the rule of living things, is one of the most important contributions ever made to knowledge, and its effects have been felt in every field of human thought.

That plants and animals constitute natural kinds, or species, had become clear by the end of the 17th century when John Ray defined them as groups of individuals that breed among themselves. In general, species were accepted like other aspects of nature as the result of special creation in each case, and there was little incentive to inquire further.

In the 18th century, doubts began to arise concerning the immutability of species, both among philosophers arguing theoretically, and among a small number of naturalists who found it difficult to distinguish between varieties of cultivated plants and domestic animals which were recognized as the diversified products of species, and species themselves which were supposed to be unchangeable. Some naturalists, including Linnaeus himself in his later years, adopted a compromise allowing that species could have descended with modifications from genera, but that genera were immutable.

With the increase in detailed knowledge of the flora and fauna of the world consequent upon the final stages of exploration, the problem of the distinction between varieties and species became acute. With a boldness and a breadth of vision amounting to genius, the French naturalist Lamarck cut the knot by proclaiming that there was no essential difference between species and varieties, that species like varieties were subject to change, and that ' transformism ', not immutability, of species was the basis of life. As it happened, there were two reasons why Lamarck's ideas were not accepted. The first was that he undertook no analysis to provide evidence for his notion of evolution; it flashed across his mind and he assumed its truth without taking the trouble to prove it. Secondly, he attempted to give an explanation of the causes of evolution which, unfortunately, raised opposition to the acceptance of the concept of evolution itself. He supposed that as a result of new needs experienced by the animal, its ' inner feelings ' or subconscious activities produced new organs which satisfied those needs. Not only was such a supposition unacceptable for the solution of the problem of the origin of species of animals, but it was totally inapplicable to plants. On the other hand, Lamarck elaborated the view originally put forward by Aristotle, and held by Maupertuis, Diderot, Buffon, and Erasmus Darwin, but which is now known to be without foundation, namely that the effects of use and disuse of an organ during the lifetime of the individual could be directly transmitted by inheritance to its offspring. There for a time the matter rested.

The Fact of Evolution

When Darwin started on the voyage of the *Beagle* in 1831, he had no reason to doubt the immutability of species. The speculations of his grandfather Erasmus counted for nothing with him because they were not supported by evidence, and those of Lamarck on the causes of evolution had the additional demerit of bringing the subject into disrepute by their fanciful nature. Furthermore, it must be added that in Lyell's *Principles of Geology*, to which Darwin owed so much for the general background of uniformitarianism in place of catastrophism that it advocated, the possibility of evolution was firmly rejected.

Three sets of observations started Darwin's revolt against the immutability of species. The first was occasioned by his studies of the fauna of the Galapagos Islands, where he found that species of finches differed slightly from island to island, while showing general resemblance not only to each other but to the finches on the adjacent mainland of South America (see pp. 4 and 74). If these species had been separately created, why should there have been such a prodigal expenditure of ' creation ' just there; why should geographical propinquity have caused these ' creations ' to resemble each other so closely; why in spite of the similarity in physical conditions between the islands of the Galapagos Archipelago and the Cape Verde Islands are their faunas totally different, the former resembling that of South America while the fauna of the latter resembles that of Africa?

The second set of observations related to the fact that as he travelled over South America, the species occupying a particular niche in nature in some regions were replaced in neighbouring regions by other species that were different, yet closely similar. Why are the rabbit-like

Figure 1. Charles Darwin at the age of 45 from a photograph taken in 1845 by Maull & Fox.

animals on the savannahs of La Plata built on the plan of the peculiar South American type of rodent, and not on that of North America or the Old World?

The third set of observations was concerned with the fact that in the Pampas he found fossil remains of large mammals covered with armour like that of the armadillos now living on that continent (see p. 64). Why were these extinct animals built on the same plan as those now living?

On the view that species were immutable and had not changed since they were severally created, there was no rational answer to any of these questions, which would have to remain as unfathomable mysteries. On the other hand, if species, like varieties, were subject to modification during descent and to divergence into different lines of descent, *all* these questions could be satisfactorily and simply answered. The finches of the Galapagos resemble each other and those of South America because they are descended from a common ancestor; they differ from one another because they are each adapted to modes of life restricted to their own particular island, one for instance feeding on seeds on the ground and another on insects in trees. The volcanic nature and physical conditions of the Galapagos Islands resemble those of the Cape Verde Islands, and yet the Galapagos birds all differ from the birds of the Cape Verde Islands. Therefore it is not the physical conditions of the islands that determine their differences. These arose because the Cape Verde Island birds share a common ancestor with the birds of Africa, whereas the Galapagos birds share a common ancestor with those of South America. The hares of South America are built on the South American rodent plan because all South American rodents are descended from a common ancestor. The fossil *Glyptodon* resembles the living armadillos because they, also, share a common ancestor; and this case is particularly important because if living species show affinity with extinct species, there is no necessity to believe that extinct types of animals have left no living descendants. They may have representatives alive today, and this means that the whole wealth of the palaeontological record of fossils is available as material for the study of the problem of evolution.

In possession of a working hypothesis that species have undergone evolution and successive origination by descent with modification from ancestral species shared in common with other species, Darwin next proceeded to search the whole field of botanical and zoological knowledge for evidence bearing on his hypothesis; for he realized that no general principle that explained the evolution of animals was acceptable unless it also applied to plants. The result was one of the most remarkable attacks on a problem ever made by the inductive method of searching for facts, whatever their import might be.

In the first place, in cultivated plants and domestic animals such as the dahlia, the potato, the pigeon, and the rabbit, a large number of varieties have in each case been produced from a single original stock. Descent with modification and divergence into several lines is therefore certainly possible within the species.

Comparative anatomy reveals the existence of plans of structure in large groups of organisms. Some parts of flowers have a structure so leaf-like that they can be regarded as modified leaves. Vertebrate animals have forelimbs which may be used for walking, running, swimming, or flying, but in which the various parts of the skeleton correspond, bone for bone, from the upper arm to the last joints of the fingers, whether the animal is a frog, a lizard, a turtle, a bird, a rabbit, a seal, a bat, or a man (see p. 70). This is what is meant by saying that such structures are homologous, and these correspondences are inexplicable unless the animals are descended from a common ancestor. Fundamental resemblance is therefore evidence of genetic affinity.

The study of comparative behaviour proves that related forms show gradations in their instincts, such as shamming death in insects and nest-building in birds. At the same time, related species inhabiting different parts of the earth under very different conditions retain similar instincts, such as the habit of thrushes in England and in South America of lining nests with mud, or that of wrens in England and North America of the males building 'cock-nests'. Why should this be, unless the different species of thrushes and wrens are descended from common ancestors in each case?

Embryology reveals remarkable similarity of structure between young embryos of animals which in the adult stage are as different as fish, lizard, fowl, and man (see p. 71). This similarity even extends to such details as the manner in which the blood-vessels run from the heart to the dorsal aorta, a plan which is of obvious significance in the case of the fish that breathes by means of gills, but not so obvious in that of lizard, chick, or man where gill-pouches are formed in the embryo but soon become transformed into different structures, and breathing is carried out by other means. This similarity between embryos is explained by the affinity and descent from a common ancestor of the groups to which they belong.

Embryology also provides evidence of vestiges of structures, which once performed important functions in the ancestors but now either perform different functions, or none at all (see p. 71). Examples of such organs are the teeth of whalebone whales, the limbs of snakes, the wings of ostriches and penguins, or the flowers of the feather-hyacinth. Since Darwin's time countless other examples have been discovered, the most striking of which are the pineal gland which is a vestigial eye, and vestiges of the egg-tooth found in marsupials although it is 75 million years since their ancestors had to use an egg-tooth to crack the shell and hatch out of their eggs. Here again, descent from common ancestral forms explains all these cases.

Knowledge of the fossil record in Darwin's time was so imperfect that nothing was then available in the way of series illustrating the course of evolution. Nevertheless, he noticed that in Tertiary strata, the lower the horizon the fewer fossils there were belonging to species alive today (see p. 68). Palaeontology therefore showed that new species had appeared, and old species become extinct, not all at the same time but in succession and gradually. Why should this be so unless new species have come into existence from time to time by descent with modification from other species?

Plants and animals are classified according to their resemblance and they are placed in one or other of a not very large number of groups, such as ferns, conifers, molluscs, or mammals. But within each of these groups, there is subdivision into other smaller groups, mammals being so subdivided into rodents, carnivores, ungulates, and primates for example. Within these again there is further subdivision, and the important point to notice is that Classification always places species in groups that are contained within other larger groups (see p. 77). This is such a commonplace that its significance is often overlooked. Why do organisms have to be classified like this? Why are they not strewn in single file up the ladder of the plant and animal kingdoms, or fortuitously like pebbles on a beach, or arbitrarily like the stars in imaginary constellations? The reason is that the arrangement of groups within groups is a *natural* classification reflecting the course of evolution. It is the result of descent from common ancestors and an indication of affinity; the differences between the groups are due to modification and divergence during such descent.

Darwin also investigated the problem of inter-specific sterility and saw that it was by no means absolute, because numerous examples can be given of different species that produce hybrids, and in some cases these hybrids are themselves fertile (see p. 75). From the point of view of breeding, therefore, such species behave like varieties. Why, then, can species not have originated as varieties, by descent and modification from other species?

From the evidence provided by all these sources, Darwin built up an irrefutable argument that species have changed and originated from other species, and that evolution has occurred. That he should have been able to do so from such few data is a mark of genius, for at the

Figure 2. Alfred Russel Wallace at the age of 46 from a photograph taken in 1869.

time when he worked out his conclusions, none of the cases had been discovered which would now be used as the most striking examples with which to illustrate the fact and the course of evolution. Chief among these are the beautiful series of fossils which reveal the evolution of the ammonites or of the horses, step by step (see pp. 65

and 67); and those which represent the precursors of the various classes and groups of vertebrates such as *Archaeopteryx* or *Pithecanthropus*.

The main steps in Darwin's proof of the fact of evolution were established by 1842 when he committed them to paper in the form of a Sketch which he expanded into an Essay in 1844 though neither was then published. Soon after this another naturalist, Alfred Russel Wallace, was led to explore similar lines of research. From some simple observations on the distribution of organisms, both geographically over the world and geologically in the fossil record, Wallace drew some equally simple conclusions that are of great importance in the history of thought which led to the realization of evolution. They show that independently of Darwin and in complete ignorance of his work, Wallace had hit upon the same solution of the problem of the mutability of species.

Wallace's observations were based on the facts, first, that large systematic groups such as Classes and Orders are usually distributed over the whole of the earth, whereas groups of low systematic value such as families, genera, and species frequently have a very small localized distribution. Secondly, 'when a group is confined to one district, and is rich in species, it is almost invariably the case that the most closely allied species are found in the same locality or in closely adjoining localities, and that therefore the natural sequence of the species by affinity is also geographical'. Thirdly, in the fossil record, large groups extend through several geological formations, and 'no group or species has come into existence twice'.

The conclusion which Wallace drew from these observations was that: 'Every species has come into existence coincident both in space and time with a pre-existing closely allied species.' Thought out about 1854, written at Sarawak in 1855, and published in the same year, Wallace's theory already allowed him to say that 'the natural series of affinities will also represent the order in which the several species came into existence, each one having had for its immediate antitype [i.e. antetype] a closely allied species existing at the time of its origin. It is evidently possible that two or three distinct species may have had a common antitype, and that each of these may again have become the antitypes from which other closely allied species were created.'

With the help of this principle, in which it is only necessary to substitute 'ancestor' for 'antitype' for the formulation of evolution to be complete, Wallace showed that it was possible to give a simple explanation of natural classification, of the geographical distribution of plants and animals including those of the Galapagos Islands, of the succession of forms in the fossil record, and of rudimentary organs which would be inexplicable ' if each species had been created independently, and without any necessary relations with pre-existing species '.

So much of the credit for the establishment of the fact of evolution has, rightly, been accorded to Darwin that it is only just that Wallace's contribution to this problem should be recognized and honoured.

The evidence on which Darwin and Wallace based their demonstration that evolution was a fact, is not only valid to this day, but has been confirmed in all the branches of science concerned as well as in many new fields. There was in their day not even an inkling of the possibilities of research opened up to comparative physiology and biochemistry, or of serology as a quantitative indicator of the amount of divergence that has taken place between related forms. Why should the chemical substance involved in the mechanism of muscular contraction in most invertebrates be arginine, whereas it is creatine in vertebrates and echinoderms, which on independent evidence are regarded as related? Why should serum immunized against man give precipitations of 64% when mixed with blood of a gorilla, but 42% with that of an orang-utan, 29% with that of a baboon, and only 10% with that of an ox? Why should syphilis attack the chimpanzee more seriously than the orang-utan, and the latter more seriously than the baboon? Why should the human ABO blood-group system also be found in the apes? The answer to all these questions is that the organisms concerned have undergone evolution from common ancestors, as a result of which members of the various lines of descent share not only structural, mental, and genetical characters, but also physiological and biochemical mechanisms, and immunological reactions.

The Mechanism of Natural Selection

Although Darwin already knew in 1837 that evolution was an inescapable conclusion to be drawn from the evidence, he did not allow himself to proceed any further with his discovery until he had found an explanation of the fact of adaptation. In a general way, all plants and animals are adapted to their environment, for otherwise they could not live (see p. 43). A man drowns in the sea ; a fish dies out of water. But there are some structures which show a particularly intimate relationship between the organism and its conditions of life. Mistletoe is a parasite that requires a tree of certain species to live on, a particular insect to pollinate its flowers, and a thrush to eat its berries and deposit its seeds on branches of the same species of tree. A woodpecker has two of its toes turned backwards with which it grips the bark of a tree; it has stiff tail-feathers with which it props itself against the tree; it has a very strong beak with which it bores holes in the tree trunk; and it has an abnormally long tongue with which it takes the grubs at the bottom of the holes. Other plants than mistletoes and other birds than woodpeckers do not have all these adaptations, and therefore, if evolution has occurred, it is necessary to give an objective explanation of how these adaptations arose.

Darwin knew that all members of a species are not identical but show variation in size, strength, health, fertility, longevity, instincts, habits, mental attributes, and countless other characters (see p. 33). He soon perceived that such variation could be, and in fact was, turned to good account by man in the course of artificial selection which he has practised in the production of cultivated plants and domestic animals since the Neolithic Age (see pp. 54 and 98). The key was selection which is the practice of breeding only from those parents that possess the desired qualities. But how could selection operate on wild plants and animals in nature since the beginning of life on earth without man or a conscious being to direct it? The solution had occurred to Darwin before he read Malthus's *Essay on Population*, but he then realized that, under the conditions of intense competition in which plants and animals live, any variations would be preserved which increased the organism's ability to leave fertile offspring, while those variations which decreased it would be eliminated (see pp. 50 and 60). In a state of nature, selection works automatically, which is why Darwin called it Natural Selection. Furthermore, it is not only individuals that natural selection eliminates but their potential offspring which, because of the fact of heredity, would have resembled them.

Darwin was then able to formulate a complete theory providing a rational explanation of the causes as well as of the fact of evolution in plants and animals. It is formally based on four propositions which he already knew to be true, and three deductions which are now also known to be true. They may be enumerated as follows:

1. Organisms produce a far greater number of reproductive cells than ever give rise to mature individuals.

2. The numbers of individuals in species remain more or less constant.
3. Therefore there must be a high rate of mortality.
4. The individuals in a species are not all identical, but show variation in all characters.
5. Therefore some variants will succeed better and others less well in the competition for survival, and the parents of the next generation will be naturally selected from among those members of the species that show variation in the direction of more effective adaptation to the conditions of their environment.
6. Hereditary resemblance between parent and offspring is a fact.
7. Therefore subsequent generations will maintain and improve on the degree of adaptation realized by their parents by gradual change.

This is the formal theory of evolution by natural selection, first announced jointly on 1 July 1858 by Darwin and Alfred Russel Wallace who had, again independently, come to the identical conclusion. It represents a step in knowledge comparable to Newton's discovery of the law of gravitation.

The Integration of Mendelian Genetics with Selection

At the time when Darwin wrote, nothing whatever was known about the laws of heredity, and all that he had to go upon was the vague notion that offspring tended to strike an average between the characters of their parents, a supposition that went by the name of 'blending inheritance', and which occasioned for Darwin the greatest difficulty with which he had to contend in formulating his theory. In the first place, if blending inheritance were true, it would mean that any new variation which appeared, even if heritable, would be rapidly diluted by 'swamping', and in about ten generations would have been obliterated. To compensate for this it would be necessary to suppose that new variations were extremely frequent. Since whole brothers, sons of the same father and mother, share an identical heredity, any difference between them would have to be due to new variation that had arisen during their own early lives, and variation would have to affect practically all members of a species. This problem of the supply of variation was a difficulty which Darwin felt so acutely that it even led him to look for a source of this supply in the supposed hereditary effects of use and disuse.

This reliance on use and disuse as a source of variation, without any effect on his main argument, is the only part of Darwin's demonstration which has had to be abandoned, and he would have welcomed the reasons for it. If only Darwin had realized it, the solution to all these difficulties was being provided by Gregor Mendel, but his results were ignored until 1900, eighteen years after Darwin's death.

The Mendelian theory of the gene, as worked out by T. H. Morgan and his colleagues with an unprecedented wealth of experimental evidence from the breeding pen and from cytological studies on the structure of the cell and its chromosomes, has established as firmly as Newton's laws of motion or the atomic theory that hereditary resemblances are determined by discrete particles, the genes, molecules situated in the chromosomes of the cells, which are transmitted to offspring in accordance with the mechanism of germ-cell formation and fertilization, and conform to distributional patterns known as Mendelian inheritance (see p. 37). The researches of C. D. Darlington and others on the structure and behaviour of the chromosomes have reached such a degree of refinement and precision that each step in the mechanism of Mendelian inheritance can actually be seen under the microscope.

The genes preserve their separate identity ; they collaborate in the production of the characters of the individual that possesses them, but they never contaminate each other ; they remain constant for long periods but, from time to time, they undergo a change known as *mutation* and involving a change in the characters which they control, after which they remain constant in their new condition until they mutate again. It has been conclusively proved that the theory of the gene applies to all plants and all animals investigated, and that the mutation of genes is the only known way in which *heritable* variation arises (see p. 37). The modifications resulting from good or bad food supply, or from the climatic conditions in which plants and animals live, are not inherited and therefore without significance in evolution.

The history of the reception of Mendelian genetics after its discovery has been peculiar. The earliest mutations discovered, often called 'sports', were usually deleterious and showed marked and discontinuous steps instead of the gradual and continuous variation which Darwinian selectionists looked for as the raw material of evolution. Selectionists therefore rejected Mendelian genetics as the source of variation. On the other hand the Mendelian geneticists, knowing that their mutations were the only source of heritable variation, thought that, as these showed wide discontinuous steps and arose suddenly, ready-made and apparently without long-continued selection, selection was inoperative in evolution and they rejected it.

With the progress of knowledge it gradually became obvious that each of the two schools of research objected to the other for reasons which were baseless. As more and more genes were identified and their effects studied, it became clear that the wide and discontinuous mutations first observed were the more easily detected extremes of a range in which the majority exert only slight effects. For the same reason, these mutations were deleterious because organisms are delicately adjusted systems, more likely to be upset by large and discontinuous changes than by small and gradual steps.

The Mendelian geneticists also had to learn two lessons. On the one hand they discovered that, although individual genes are associated with particular characters, their control of those characters is also affected by all the other genes, which constitute an organized gene-complex. As a result of previous mutations, gene-complexes of plants and animals in nature contain many genes and these are sorted out and recombined at fertilization in astronomically numerous possibilities of permutations. These recombinations have been shown to bring about *gradual and continuous* changes in the characters under the major control of individual genes. Sir Ronald Fisher demonstrated the significance of this by showing that a mutant gene that now exhibits the quality known as dominance has *gradually become* dominant from a previous intermediate condition. This is what has happened to those mutations that confer a benefit on their possessors, and in their case there has been a selection of gene-complexes in favour of those which accentuate the effects of a favourable mutant, so that these effects are manifested even if the mutant gene in inherited from only one parent, which is the definition of dominance. Conversely, with genes that place a handicap on their possessors, there has been a selection of gene-complexes in favour of those which suppress the effects of such genes so that they are manifested only when the mutant gene is inherited from both parents and the gene has *gradually become* recessive; or suppress them even further when the effects of such a gene are obliterated and the gene becomes what is known as 'modifier' without major control over characters. It has even been demonstrated by E. B. Ford under rigorous experimental conditions that one and the same mutant gene can be made to *become* dominant in one strain and recessive in another simply by selecting as parents those individuals whose gene-complexes accentuate or diminish the effects of the gene (see p. 42).

The Mendelian geneticists' second lesson was the realization that, although the effects of the mutations which they first observed appeared to be clear-cut, they were already the results of selection in past gene-complexes. For these mutations have occurred before, and the gene-complexes have become adjusted to them. The fact that a single gene may now act as a switch controlling the production of one or other character-difference does *not* mean that this character-difference originally arose at one stroke by one mutation of such a switch-gene, because it has probably been built up gradually as a result of past selection in the gene-complex.

It is therefore clear that mutations and recombinations of genes provide the supply of variation on which selection acts to produce evolution exactly in the way Darwin's theory requires. Its requirements are exacting, for, as T. H. Huxley pointed out, some organisms have evolved slowly and others have evolved fast, and he saw that natural selection was the only mechanism that could satisfy both those requirements. It is able to do so because Mendelian inheritance is capable of producing both diversity and stability. As E. B. Ford has said, an immense range of types must be available for natural selection to act upon, and this is provided by mutation and recombination of genes ; yet when a favourable gene-complex has been achieved it must not be dissipated and broken down, and this is provided against by the facts that the genes do not blend or contaminate one another, and that they mutate only rarely.

The Significance of Particulate Inheritance in Evolution

The particulate theory of inheritance which Mendelian genetics has established involves a number of consequences of fundamental importance for the problem of evolution (see p. 41). In the first place, the substitution of this quantitative and deterministic science for the vague and baseless notion of ' blending inheritance ' completely disposes of the difficulty under which Darwin laboured to account for the necessary supply of variation on which natural selection could act. The most characteristic feature of the Mendelian gene is that it *never* blends, but retains its identity and properties intact for long periods of time until it mutates, after which it remains intact in its new condition until it eventually mutates again. This means that the amount of variation, or *variance*, present in a population resulting from previous mutations, is not only *conserved* through generation after generation, but is actually *increased* as a result of the recombinations of the gene-complexes in their innumerable possible permutations. This power of increase is one of the most important results of the bi-parental method of reproduction and is the reason why organisms possessed of this mechanism have evolved further than those that lack it (see p. 39).

This conservation of variance is to be considered in relation to the rate at which mutation normally occurs. It has been calculated in organisms as diverse as a bacterium, a maize-plant, a fruit-fly, and in man, that any given gene mutates in about one in half a million individuals. It is also clear that this rate is itself the result of selection, and that although seemingly slow, it has been adequate to provide the requisite basic heritable variation which the mechanism of germ-cell formation and fertilization has multiplied, and on which selection has worked to produce whatever evolution has taken place. In other words, mutation not only need not, but must not be more frequent than a slow rate, which is enormously more infrequent than what it would have been if ' blending inheritance ' were a fact, and Darwin's difficulty in accounting for an adequate supply of variation is removed.

As the originating mechanism for basic heritable variation, mutation has naturally been the subject of intense study. It has been found that certain physical and chemical factors including radio-activity can accelerate the rate at which mutation would naturally occur, but that these induced mutations are similar to those which occur and recur normally. Even if certain mutagenic agents may induce particular mutations, no correlation exists between the agents and any quality of usefulness of the mutations. Mutations take place blindly, as H. J. Muller has said, subject only to the restriction of chemical possibilities in the gene-molecules. This shows that there is no basis for attempts to explain the origin of heritable variation by appealing to environmental factors to evoke appropriate adaptive inherited responses. Nor is there any basis for the view that the environment would evoke appropriate inherited responses if its actions were continued for a sufficient length of time, because, as J. B. S. Haldane showed, such responses as might be significant in evolution would be detected within the period of the experiment carried out.

In organisms that reproduce by simple division of the whole body such as bacteria, special conditions apply because reproduction in them involves not only transmission of genetic material in the form of genes, but also transmission of bodily characters, since the latter are carried over wholesale from ' parent ' to offspring. In bacteria it is possible for organic molecules such as bacteriophage particles to enter organisms and become incorporated in the genetic mechanism so as to behave like genes. In higher organisms it may perhaps be possible for genes from grafted or injected tissues to do likewise. These results are full of promise as a field of research into the nature of genes and of mutations, but they do not in any way invalidate the principles of Mendelian genetics and inheritance.

Mutations are chemical changes in the gene-molecule, and since chemical stability is not absolute, the puzzle about mutations is not so much that they occur as that they occur so infrequently. This ignorance of the causes which determine the directions in which mutations take place, if such causes indeed exist, is, strange to relate, no handicap to the understanding of the mechanism of evolution, because *it is selection, not mutation, that determines the direction of evolution.* This all-important conclusion is based not only on detailed experimental studies on the effects of selection in nature, but also on the demonstration by Sir Ronald Fisher of a general principle. The effects of selection in changing the frequency of genes in a population have been calculated for various percentage benefits in survival-value conferred by such genes. At the estimated natural average mutation-rate of one in half a million, no mutant gene has the slightest chance of maintaining itself against even the faintest degree of adverse selection. Furthermore, if the direction of evolution were determined by the direction of mutation, it would be necessary to suppose that such mutations must be predominantly favourable. But the majority of mutations were unfavourable when they first arose, and natural selection has acted against them by converting the resulting mutant genes into recessives or suppressing them into the condition of mere modifiers ; or their effects have been improved by a changed gene-complex resulting from natural selection, and the genes are then preserved, and become dominant (see pp. 13, 42). It is natural selection, not mutation, that has governed the direction as well as the amount of evolution, and it has been estimated that if mutation were to stop now, there is already sufficient variation in the plant and animal kingdoms for evolution to continue for as long in the future as it has gone on in the past up till now.

The bearing of this demonstration on hypotheses that attempt to explain evolution by postulating the existence of agencies capable of directing mutation is plain. It means that all such theories as invoke the effects of use and disuse, 'inheritance of acquired characters', environmental stimuli, 'organic selection', 'inner feelings', 'inherited memory', momentum along particular directions, orthogenesis, nomogenesis, and others, which assume that mutation can be made to follow adaptively desirable directions, are not only devoid of any known mechanism by which the direction of mutation might be brought about, and devoid of evidence regarding the existence of such mechanisms, but they involve a cause 'which demonstrably would not work even if it were known to exist'. It is therefore not surprising that in spite of repeated attempts, many undertaken with impure and insufficiently standardized genetic material, and others in which the results were simply faked, no evidence has been provided that the effects of use and disuse or adaptive response to environmental conditions are inherited or induce appropriate mutations. From the evidence provided by genetics, natural selection is the only mechanism capable of explaining evolution.

Natural Selection, 'Improbability', and 'Chance'

An argument sometimes used against the efficacy of natural selection involves the claim that the initial stages in the evolution of complex structures or functions could not have been favoured by natural selection until such structures or functions had reached a certain level of perfection. Like all other arguments of the *non possumus* type, this one melts away before the progress of knowledge. A case in point is that of the electric organs of fish, developed out of muscles which are capable of discharges strong enough to catch prey and defend the fish against its enemies. These organs are clearly adaptive and confer survival-value on their possessors, but the question arises what functions they could perform in the initial stages of their evolution when it must be supposed that their power was weak, too weak to kill prey or to deter predators. Darwin himself was well aware of this problem, and he met the argument by pointing out that 'it would be extremely bold to maintain that no serviceable transitions are possible by which these organs might have been gradually developed'. He has been proved to be right, because of the discovery by H. W. Lissmann that weak electric discharges given off by certain fish function in a manner analogous to radar equipment, and serve to convey information of the proximity of objects in the water. Electric organs can therefore be adaptive even when they are too weak to kill prey or deter predators.

Another case may be cited because it illustrates the manner in which an adaptive result may be achieved without itself being a direct object of selection. Colour-vision has been evolved independently in many groups of animals. Among the light-sensitive elements in the eye, some are specially sensitive in dim illumination ; others confer acuteness of vision in bright light when they are individually innervated, with the result the light-stimuli are perceived separately by very small areas of the retina. In each of the two functions of seeing in the dark and seeing accurately in the light, increased efficiency confers survival-value from the very start of the improvement. But when both these functions have been achieved in the same eyes, as E. N. Willmer has indicated, a mechanism is produced in which the visual elements are differentially sensitive to light of different wave-lengths, and this is the basis of colour-vision. The emergence of colour-vision as an unexpected ' bonus ' resulting from the perfection of two other functions is a concrete example of the principle to which Lloyd Morgan applied the term ' emergent evolution '.

The fact that an organism is adapted to the conditions of its environment has been used as an argument in favour of the view that such adaptations have been directly caused by the environment in response to stimuli. These stimuli can, in many cases, evoke an adaptive response from an organism during its own life, as when friction thickens the skin of man's hands. But such responses are not inherited, and it is therefore necessary to regard adaptations as the result of random variations channelled by natural selection into directions which are useful and thereby give the appearance of having been purposeful in origin.

There is an additional way in which selection might result in production of adaptations. When an organism responds to an environmental stimulus, its capacity to make the response is due to its possession of the relevant genes. C. H. Waddington has shown that, in some cases, the effects of these genes can be modified by changes in the gene-complex as a result of selection. The capacity to respond to environmental stimuli can, in this way, be decreased or increased, and even increased to the point where the response is evoked by an inherited modification in the process of development of the organism which then provides a stimulus which has become substituted for the original environmental stimulus no longer required. It is possible that this principle, known as ' genetic assimilation ', may be of wider application though it has not yet produced anything adaptive.

It has also been objected that natural selection is a difficult concept to apply to the evolution of very complex adaptations involving co-ordinated variations either in one and the same organism, or even in two different organisms. It is not necessary to go far afield to find examples of this, for in all animals with separate sexes and internal fertilization, there has been a separate yet harmonious evolution of the reproductive organs in the two sexes. It has been supposed that such situations argued so high a degree of 'mathematical improbability' that they could not be explained as a result of natural selection, which was, very erroneously, called 'chance'. To this attitude there are several answers.

In the first place, although variation is haphazard and mutation is random, it is neither variation nor mutation but *selection* that determines the course of evolution (see p. 16). The operation of natural selection is not a game of luck but is governed by rigid conditions and objective causes that leave no room for 'chance'. It is therefore meaningless to refer to evolution by natural selection as the result of 'blind chance' or the accumulation of 'lucky accidents'.

In the next place, those who invoke mathematical improbability against natural selection can be refuted out of their own mouths. H. J. Muller has estimated that on the existing knowledge of the percentage of mutations that are beneficial, and a reasoned estimate of the number of mutations that would be necessary to convert an amoeba into a horse, based on the average magnitude of the effects of mutations, the number of mutations required on the basis of chance alone if there were no natural selection, would be of the order of one thousand raised to the power of one million. This impossible and meaningless figure serves to illustrate the power of natural selection in collecting favourable mutations and minimizing waste of variation, for horses *do* exist and they *have* evolved.

It is worth while to study the question of improbability more closely. As Sir Ronald Fisher has pointed out, improbability bears a different aspect when considered from time before or time after the event. The probability that any man alive today will have sons, grandsons, and successive descendants in the male line uninterruptedly for one hundred generations, is infinitesimally small. Yet every man today is the living proof that this contingency, so highly improbable, as it may have seemed one hundred generations ago, has nevertheless occurred. Similarly, the effects of natural selection are the reverse of chance when considered *ex post facto*; they are rigorously determined and what they have done is to channel random variation into adaptive directions and thereby simulate the appearance of purposive change. This is why natural selection has been paradoxically defined as 'a mechanism for generating an exceedingly high degree of improbability'.

Mention of purpose introduces the notion of teleology or fulfilment of design, which has sometimes been invoked to explain the production of complex adaptions. Teleology and providential guidance are double-edged weapons with which to attack the problem of evolution, because it can be shown that the more detailed the adaptation, the more 'improbable' it may appear as a product of 'chance', the more likely its possessor is to be doomed to extinction through inability to become adapted to changed conditions, when conditions change.

Structures may be developed which at first benefit individuals in their competition to survive; but by continued selection such structures may become exaggerated and lead to the extinction of the species. This seems to have been what happened to the Huia-bird where mated pairs constantly remained in company together, and the beaks of the male and female reached an extraordinary disparity of size in adaptation to their very special feeding (see p. 46), but failed to enable the birds to obtain ordinary food when their special diet was unavailable. Excess, even of adaptation, is harmful, and the fossil record shows that the vast majority of lines of evolution have led to extinction, which is a grim comment on the alleged powers of providential guidance and purpose.

From the undoubted fact that many of the products of the plant and animal kingdom convey to man the aesthetic quality of beauty, it has been supposed that beauty is an end in itself to which the criterion of usefulness and survival-value could not be applied, and therefore that it could not be imagined as a product of evolution. To this argument, Alfred Russel Wallace opposed the demonstration that if the quality of beauty were an exception to the principle of evolution by natural selection, it would be necessary to find an explanation for the existence of so much in plants and animals that is positively ugly.

Darwin showed it to be an invariable rule that: 'When a flower is fertilized by the wind it never has a gaily-coloured corolla.' The beauty of flowers has been gradually achieved because of the survival-value (of cross-fertilization consequent upon the attraction of insects to such flowers) conferred on plants possessing them. The beautiful colours and structures of birds and some other animals have resulted from the survival-value conferred on successful competitors in sexual selection.

This demonstration of what may be called the natural nature of beauty has been developed still further by Ray Lankester in the course of a soliloquy on alpine flowers: 'All beauty of living things, it seems, is due to Nature's selection, and not only all beauty of colour and form, but that beauty of behaviour and excellence of inner quality which we call "goodness". The fittest, that which has survived and will survive in the struggle of organic growth, is (we see it in these flowers) in Man's estimation the beautiful. Is it possible to doubt that just as we approve and delightedly revel in the beauty created by "natural selection", so we give our admiration and reverence, without question, to "goodness', which also is the creation of Nature's great unfolding?'

In many of the higher animals, parental care and self-sacrifice, in the interest of other members of the family such as incubating or gravid females and young, have been favoured by natural selection and conferred benefit on the species. From earliest human times, the survival value of altruistic behaviour has been enhanced because of the prolongation of childhood and the consolidation of the family that have characterized the evolution of Man. The size of the unit within which altruistic behaviour conferred survival-value has grown progressively larger, but fitfully, as history and anthropology have shown, from the family to the clan, the tribe, and the nation. In this manner, ethical standards of conduct and morality have arisen which can be seen to develop in individuals and have been seen to evolve in societies. Between these units, competition on the sub-human level of natural selection has persisted. With the development of man's higher mental faculties, the pooling of experience by speech-communication and its storage in memory, conscious choice and purposiveness became factors in evolution. Hence his subsequent psycho-social evolution has been of a nature different from that of other organisms, because it was no longer governed solely by natural selection and therefore involved processes outside the scope of natural selection.

Natural Selection in Action

Natural Selection can be seen to be at work here and now in directing evolution. Modern techniques of study of genetics in populations in the field, developed by T. Dobzhansky and E. B. Ford, have shown that the relative longevities of variants in different environments can be directly measured, and that the effects of such differential mortality have been to produce evolutionary change. An example of this type of research is that of H. B. D. Kettlewell on 'industrial melanism' in moths (see p. 49). Up to 1848 the British Peppered Moth existed in its typical grey form known as *Biston betularia*, which is remarkably well adapted to resemble the lichens on the bark of trees. From that date, a dark melanic variety appeared, known as *carbonaria*, which is extremely conspicuous against the natural bark of trees. It is controlled by a single dominant Mendelian gene and is slightly more vigorous than the normal grey type. Nevertheless, because of its conspicuous colour the *carbonaria* variety was constantly eliminated, and this variety only persisted in the populations of the Peppered Moth because the same mutation kept on occurring again and again. The Industrial Revolution brought about a marked change in the environment, since the pollution of the air by increasing quantities of carbon dust killed the lichens on the trees and rendered their trunks and branches black. Under these conditions it is the *carbonaria* variety which is favoured and the *betularia* penalized. This has been proved by direct observation of the feeding of birds, and by measurement of the survival rates of the different forms in the different environments. The dark *carbonaria* form survives 17 % less well in an unpolluted area and 10 % better in a polluted area. One hundred years ago the dark variety of the Peppered Moth formed less than 1 % of the population ; today in industrial areas it forms 99 %, and selection has made it more intensively black than when it first appeared.

The case of melanism in the Peppered Moth also introduces a principle to which L. Cuénot drew attention and gave the name of ‘ pre-adaptation ’. The melanic form of the Peppered Moth happened to be ‘ pre-adapted ’ to conditions which were only subsequently realized ; or, in other words, if the industrial revolution had not taken place, the melanic variety would never have become adaptive at all, and would have suffered the same fate as the countless other mutations resulting in variations which, whether ‘ pre-adapted ’ or not, have been eliminated because they fell short of the requirements imposed by natural selection.

The evolutionary change actually witnessed in the Peppered Moth is directly attributable to selection, and it is matched by similar studies on other forms. Experiments by A. J. Cain and P. M. Sheppard on the survival rate of snails with shells of different colours and banding patterns living on dark or light coloured backgrounds, have shown that selection does not act like an all-obliterating steam-roller going in one direction. As the seasons change, the adaptive value of the colour of a shell changes from disadvantageous to advantageous and back again, which proves that the effects of selection vary from place to place and from season to season, and that the balance between an organism and its environment is delicate, changing, and dynamic (see p. 60).

The phenomenon of Batesian mimicry has also been proved not only to be adaptive and to confer survival-value, but also to have been achieved by selection (see p. 52). E. B. Ford has shown that the degree of perfection with which the mimics copy their models is a function of the prevalence of the models. At places like Entebbe where models are numerous, the percentage of imperfect mimics in the populations of *Papilio dardanus* is only 4. At Nairobi, on the other hand, where the models are 70 times less numerous than at Entebbe, the imperfect mimics are 8 times more numerous and constitute 32 % of the population. Less survival-value is conferred by resemblance to a model when it is too infrequent to teach predators to shun it, and there is then less selection-pressure on the mimic to resemble it.

While the overriding importance of the effects of selection are now generally realized, it has been suggested that when populations are split up into very small isolated colonies, changes in the relative frequencies of different genes might result from the errors of random sampling in the formation of the germ-cells and their fertilization, without involving selection. This concept, advanced by Sewall Wright and known as ‘ random genetic drift ’, has been invoked as a possible cause of non-selective, non-adaptive evolution. It has however been invalidated by the results of experimental studies in the field such as those of Sir Ronald Fisher and E. B. Ford on moths, which have shown that selective factors are much more important than casual non-adaptive factors in determining the relative frequency of genes and bringing about close adaptation to local environmental conditions. Even in comparatively numerous populations, from one generation to the next there are fluctuations in gene-ratio larger than can be attributed to random sampling and which are controlled by selection. Such effects as may be due to random sampling in small populations can only be of negligible significance in evolution.

Selection frequently works on a basis of compromise. Among the natives of Africa there is a condition known as sickle-cell in which the red blood-corpuscles are deformed and shaped like the blades of sickles. This is controlled by a Mendelian gene which, when inherited from both parents (homozygous), produces an extreme effect which frequently kills the subject by thrombosis. When inherited from only one parent (heterozygous) the danger from thrombosis is not so great. In areas where malaria is present, however, there is a positive advantage in possessing the sickle-cell gene, because it produces abnormal haemoglobin unsuitable for the malaria parasite. In accordance with the prevalence of malaria in the environment, therefore, a balance is automatically struck in the population between the danger of dying from malaria if the individual has no sickle-cell gene, and the danger of dying of thrombosis if the individual has two sickle-cell genes. Survival-value and ability to leave more offspring therefore accrues to the possessors of one sickle-cell gene up to a certain frequency, and this example shows in what unexpected ways selection is able to make the best even of a bad gene-complex (see p. 48).

Natural Selection and Palaeontology

The palaeontological record provides evidence of the course of evolution in the past. Fossil material is in places now so rich that it can be used for quantitative studies in evolution. First, estimates of time based on disintegration of radio-active material enable various levels of evolutionary lineages to be dated and the time measured during which certain changes have occurred, thereby providing quantitative evidence of evolution-rates and of the duration-times of genera and species. Statistical study of large samples of fossil materials enables the variability of the different species to be assessed. By such methods, G. G. Simpson calculated that the evolution of the horse from *Hyracotherium* to *Equus* occupied 60 million years and involved passage through 8 genera each of an average of duration-time of 7·5 million years, 30 species of a duration-time of 2 million years each, and 15 million generations each reaching maturity in 4 years. These data can be compared with those obtained from other groups of animals in which they differ considerably. The results show that evolution-rate is not correlated with variability, nor with generation-time, and that selection controls the direction and the intensity of evolution.

These results are all the more important because in the past, some palaeontologists, unequipped with knowledge of modern genetics, have imagined that from tracing the *course* of evolution in the lineage of fossils which they established, they were in a position to draw conclusions about the *cause* of such evolution. Some have thought that they had found support for the inheritance of acquired characters although they knew nothing about inheritance ; others imagined that as the lineage of some fossils showed linear progression of certain characters, they were justified in concluding that evolution involved an innate directional component, an expression of ' momentum ' leading to evolution in ' straight lines ' which they called ' orthogenesis '. They failed to realize that, if selection in a particular direction benefits an organism, continued selection in the same direction will, up to a certain point, benefit it further. Others again have concluded from their materials that a distinction in principle could be made out between ' big ' evolution leading to large evolutionary changes, and ' small ' evolution producing trifling results. None of these speculations can stand up to the *evidence* that selection determines the course of evolution, its speed or its slowness, the magnitude or pettiness of the effects produced, and its direction which if constant for any length of time simulates ' orthogenesis '.

The detail in which the structure of many fossil animals is now known enables conclusions to be drawn regarding the manner in which one type evolved into another. The fossils representing the transitional forms between fishes and amphibia (*Ichthyostega*), amphibia and reptiles (*Seymouria*), reptiles and birds (*Archaeopteryx*), reptiles and mammals (Ictidosaurs), all agree in showing a mosaic of characters, some of which are completely typical of the group from which evolution was taking place, while other characters are equally typical of the group towards which evolution was taking place. In *Archaeopteryx*, for instance (see p. 70), the tail is completely reptilian while the feathers are completely avian, and no characters appear to be intermediate. Since these animals resemble a mosaic in which the various pieces can be changed separately, the term ' mosaic evolution ' has been applied to this mode of transformation.

An objection to natural selection has been raised on the grounds that as species differ in many characters, it was difficult to see how several parts of an organism could be modified at the same time. Darwin countered this objection by showing that ' there is no necessity for supposing that all parts of any being have been simultaneously modified ', and mosaic evolution provides examples of changes affecting the various parts one by one.

Comparison can be made between young and adult stages of ancestral and descendant forms. For a time it was supposed that the adult ancestor was reflected in the young descendant in accordance with the now abandoned theory of recapitulation which was based on erroneous conceptions and inaccurate data. On the other hand, numerous cases have been found in which the adult descendants resemble the young ancestors, as a result of retardation in development and retention of youthful characters in the adult stage. This mode of evolution which is known as paedomorphosis, or ' Peter Pan evolution ' because it leads to descendants that do not ' grow up ' like their ancestors, is associated with the production of important groups such as insects, vertebrates, and man (see p. 72). Natural selection has all stages in the life-history to choose from in selecting the best-adapted type for each species.

There are several groups of animals in which the young forms, subjected to their own particular conditions of natural selection, underwent an evolution on their own without affecting the structure of the adults. Examples are the caterpillars of moths and the larvae of starfish. But young forms have soft bodies, and they therefore failed to become preserved as fossils. In such cases the evolution of the young was 'clandestine' from the point of view of palaeontology, and the effects of such evolution only became preserved as fossils when, by paedomorphosis, the organization of the ancestral young became that of the adult descendant. Gaps in the fossil record are therefore to be expected in the ancestry of groups that have evolved by paedomorphosis.

It is of interest to consider how far it is possible to extrapolate the results of modern genetics into the palaeontological past. C. R. Diver has shown that in snails the patterns of banding found today were already in existence in Pleistocene times. It is necessary however to beware of concluding that because characters are similar they must be controlled by the same genes. Even in one and the same species today, the gene-complex can undergo permutations which reproduce the same structures with different genes. An example is provided by the gene ' eyeless ' in *Drosophila* which produces flies with very small or no eyes and is, of course, extremely harmful. Eyeless flies can, however, be made to breed, and although the mortality is very high, progeny can be reared which after a few generations have eyes like normal flies. In such a stock the ' eyeless ' gene is nevertheless present unaltered, as can be proved by mating the flies with normal flies, when the effects of the ' eyeless ' gene manifest themselves in all their force, in the second generation because 'eyeless ' is recessive. This result is therefore in perfect accordance with the principle that Mendelian genes do not become contaminated. What has happened during the inbreeding of ' eyeless ' flies is that the reshuffling of the other genes has produced a gene-complex in which the harmful effects of 'eyeless' have been suppressed.

The gene-complex is therefore a dynamic system, as S. C. Harland concluded from his researches on Cotton. Genes compete, i.e. are selected in the gene-complex, old genes being dropped and new genes incorporated. During the course of evolution the effective membership of the gene-complex must have changed, and it is not legitimate to conclude that because a character or a structure like the eye of vertebrates was in existence 400 million years ago, it was then controlled by the same genes as control it now. The evidence is entirely opposed to such a static view, and it is precisely because the gene-complexes change that characters, structures, and organisms have evolved.

The New Systematics and the Origin of Species

The researches on industrial melanism in the Peppered Moth, banding and colour of snails, mimicry in butterflies, local adaptation in moths, and sickle-cell in man, which have here been briefly described, are examples of new techniques of experimental studies on evolution in the field that has grown out of what Sir Julian Huxley has aptly called 'The New Systematics', to which he has himself so much contributed. Systematics, the study of species and of the higher groups of classification, began by the recognition of differences between species, defined from type-specimens preserved in museums. But with the realization that species now or in the past are or were populations of live plants and animals in nature, living under varying conditions of equilibrium with each other and the inorganic factors of their environments, themselves showing geographical variation in space, and undergoing variation in time, subject to mutation and recombination of their genes, constantly under the influence of selection, species can no longer be considered as static milestones of evolution, for they are themselves the dynamic systems by which the roads of evolution are trodden. As genes mutate and are reshuffled, and geographical races invade new ecological niches, advance and retreat, it is already possible on a map to mark out lines of 'gene-flow', as R. C. Stebbins has suggested from his researches on Californian newts. It may become possible to plot the areas of gene-complex alteration, as can to a certain extent already be done for the origin of cultivated plants such as wheat; but such maps will be continually changing, like the species themselves.

Nobody would have welcomed these developments of biological

science more than Darwin himself, as a glance at the last few pages of *The Origin of Species* will show. It is therefore appropriate to return to the problem with which this book began. As is now certain, species are not immutable but have undergone change, and many examples have been given above. Evolution can take place up to a point without the production of new species, but if this process continues the time must come when new species originate, and it is legitimate to ask whether modern research has revealed any evidence of this. The answer is that new species can be seen originating in nature here and now, and new species have been artificially produced in the laboratory.

Speciation takes place when, for various reasons, populations cease to breed with neighbouring populations and, under different conditions of selection, accumulate heritable variations by mutation and recombination of genes in different directions. As E. Mayr has shown, some form of biological isolation between portions of populations is a necessary condition for divergence leading to the formation of new species and higher groups (see p. 73).

Among the kinds of isolation that are chiefly responsible for the origination of species, geographical isolation is the most important, and involves physical barriers such as oceans, mountain-ranges, or deserts which separate whole populations. Geographical races are the chief raw materials from which new species are formed, and it was the different finches on the different Galapagos Islands which first suggested to Darwin that evolution had occurred (see pp. 4 and 74). Here, to various extents, geographical isolation has assisted the origination of a number of species.

A case in which geographical isolation may be expected to produce its effects at almost any moment now is provided by the gulls (see p. 73). These birds occupy a zone shaped like a ring round the North Pole and form what B. Rensch has called a chain of races. Starting with the British lesser-black-backed gull with its dark mantle and yellow legs, this is found to grade into the Scandinavian lesser-black-backed gull, and, continuing in an easterly direction around the chain, this in turn grades into the Siberian Vega gull with its lighter mantle and dull flesh-coloured legs. The Siberian gull grades into the American herring-gull which, in turn, grades into the British herring-gull with its light mantle and pinkish legs. Although the British lesser-black-backed gull may be regarded as belonging to the same species as all the other gulls in the chain to the east of it, when it is compared with the other end of the chain represented by the British herring-gull, the two may almost be regarded as separate species. Already they differ not only in colour but in habits, for the latter nests on cliffs and is dispersive in winter whereas the former breeds inland on moors and is migratory in winter. If at any time the chain is severed by the erection of a sterility barrier at any point, either through inability to breed, or through a rupture of the chain by local extinction of the gull population, the two British gulls will then belong to independent species.

Geographical isolation is important for the origin of species of plants as well as of animals; but there is another form of isolation which is largely restricted to plants and involves the sudden erection of sterility barriers between individuals in the same population as a result of changes in the chromosome mechanism. This is known as genetic isolation. When *Primula verticillata* is crossed with *Primula floribunda*, hybrid offspring are produced, but they are sterile because the chromosomes of one parent species are incompatible with those of the other, and the intricate machinery involved in the formation of germ-cells is thrown out of gear. Occasionally, however, the hybrid plant undergoes doubling of its chromosomes, a condition known as polyploidy, and when that has occurred the hybrid is able to breed with hybrids similar to itself because all the chromosomes have compatible partners, but it is sterile with both parent-species. Furthermore, the hybrid is not only true-breeding but is different in structure and in habit from each of its parent species. It therefore fulfils all the criteria of a species and has been called *Primula kewensis* (see p. 75). Many other new species have originated in this way from intentional hybridization followed by accidental polyploidy. Some of these artificially produced species have been found to be identical with, and to breed with, wild species, and this is the proof that this method of species-formation occurs in nature.

There is one further point which calls for attention. Darwin called his famous book *The Origin of Species* and the title was very appropriate because plants and animals occur in populations known as species, and he was concerned to show how species become different from other species, and therefore how new species arise. New species arise by evolution, not by special creation; but evolution must not be regarded as synonymous with the origin of species. Evolution defined as descent with modification can take place to a certain extent without involving the origination of a new species, and an example of such evolution is provided by the substitution of the melanic variety of the Peppered Moth for its light-coloured form in certain environments (see p. 49).

There are cases in which an originally uniform population becomes split into two portions which do not interbreed because of the erection of a sterility barrier between them of an ecological or genetic nature (see p. 75), and these two portions may have to be regarded as incipient different species, even although they may not be distinguishable in appearance. Such species are called Sibling species.

New species originate as a result of a process of isolation between portions of a population, and in these portions, exposed to different conditions of natural selection, different variations are selected, resulting in progressive divergence in structural and physiological characters and in reproductive incompatibility.

Since different species differ not only in the precise conditions of their origins but also in the times since they originated, they also differ in the degrees of difference which distinguish them from other species. This is the explanation of the curious fact that, although species are the units in which all living organisms are grouped, the status of what is called ' the species ' is very difficult to define.

There is first the criterion of structure: members of a species visibly resemble each other more closely than they resemble members of any other species. The criterion, which is frequently the only one that can be applied, as for instance to preserved specimens and fossils, suffers from the criticism that the attribution of specific rank to a population is an arbitrary act at the discretion of the taxonomist, and this has led to the reproach that such species are subjective concepts and have no existence outside the mind of the scientist who ' made ' them.

Such a criticism cannot be levelled at the biological criterion of the species, which is that of a group of organisms reproductively isolated from all other organisms. Whether a population does or does not interbreed with another population is a matter of fact which can be proved or disproved by observation and experiment on living populations in nature; and a species defined in this manner is not arbitrary but has objective reality and existence.

The corollary to the biological criterion of reproductive isolation surrounding a species is that of fertility between the various populations within the species, even if these populations are distinguishable geographical races. This is the concept of the species as a ' gene-pool ', of such a kind that an hereditary character of any member of the species may potentially be transmitted to any other individual of a later generation in any part of its range : a process called ' gene-flow ' as mentioned above (see p. 27).

The species is difficult to define because species are not necessarily the end-products of finite changes, but stages in a continual process of potentially infinite change in which they may be at any level, as H. W. Parker has pointed out. Like stars at different stages of their evolution, species must be the object of research to determine their status in each case.

The Centenary of Evolution by Natural Selection

In conclusion it may be said that during the hundred years that have elapsed since Darwin and Wallace first published their theory, the fact of organic evolution is now universally accepted and its mechanism has been explained.

The alternative to evolution is so naïve that it comes as a shock to realize that as recently as one hundred years ago, ideas such as called for the following questions could still be current : ' Do they really believe that at innumerable periods in the earth's history certain elemental atoms have been commanded suddenly to flash into living tissues? Do they believe that at each supposed act of creation one individual or many were produced? Were all the infinitely numerous kinds of animals and plants created as eggs or seeds, or as full grown? and in the case of mammals were they created bearing the false marks of nourishment from the mother's womb? ' Darwin might well allow himself to ask these questions, for he and Wallace had found the answer to them.

So soundly was the theory of evolution by natural selection grounded that research does nothing but confirm the links in its chain of evidence and the inferences to be drawn from them. Its field has extended from the explanation of the production of plants and animals to every aspect of the intellectual life of man, and it would be imprudent to doubt that its greatest triumph may yet lie in the highest aspect of that life. Some persons have attempted to discredit natural selection on the grounds that being a destructive agent it cannot produce anything new or make the fit fitter. Such persons have only demonstrated that they have neither understood the problem nor studied *The Origin of Species*, in which Darwin carefully pointed out that ' several writers have misapprehended or objected to the term Natural Selection. Some have even imagined that natural selection induces variability, whereas it implies only the preservation of such variations as arise and are beneficial to the being under its conditions of life.' Variation produces novelties at random, but selection determines which are preserved. Only a genius could have discovered a key of such simplicity to so great a problem. Only ignorance, neglect of truth, or prejudice could be the excuse for those who in the present state of knowledge, without discovering new facts in the laboratory or in the field, seek to impugn the scientific evidence for evolution.

With such new formulations as may be required, the concept of evolution by natural selection continues and will continue to provide what Darwin hoped when he wrote in 1837 in his Notebook: 'My theory would give zest to recent and fossil comparative anatomy; it would lead to the study of instincts, heredity and mind . . . to closest examination of hybridity—to what circumstances favour crossing and what prevents it—and generation, causes of change in order to know what we have come from and to what we tend. This and direct examination of direct passages of structure of species, might lead to laws of change, which would then be the main object of study, to guide our speculations.'

With the same confidence as it accepts Copernicus's demonstration of the movement of the earth round the sun and Newton's formulation of the laws of this movement, science can now celebrate the centenary of the first general principle to be discovered applicable to the entire realm of living beings.

THE EXHIBITS ILLUSTRATING EVOLUTION

In such a field as that of Evolution in which the observations and experiments that have been made during the past hundred years are beyond computation, it is naturally impossible to present more than the barest framework of selected examples to illustrate its general principles. They have been chosen so as to indicate the basic evidence bearing on the following subjects: Variation, Adaptation, Selection, the Fossil Record, Morphology and Embryology, Geographical Variation, Isolation, the Formation of new Species, Classification, and the Evolution of Man.

The exhibits are described in sequence, beginning at the First Bay on the East Side of the Central Hall, and ending at the Fourth Bay on the West Side. Unless this sequence is adhered to, the visitor may find that he has missed certain steps in the argument, which is as follows: The *Evolution* of plants and animals has been brought about by the occurrence of *Variation* due to *Mutation* and *Recombination* of Mendelian genes, acted upon by *Natural Selection*, resulting in changes in portions of populations *isolated* by geographical or other factors, with gradual divergence until the differences between the populations prevent them from interbreeding, when they become separate species.

The causes of *Evolution* are therefore *Inherited Variation*, *Selection*, and *Isolation*, and its results are *Adaptation* and *Species-formation*.

East side: first bay
Right-hand case

Variation

Plants and animals occur in populations of interbreeding individuals constituting Species. While species have definable characteristics of form and life-history, the populations constituting them are not composed of identical individuals, but show *Variation* (see p. 10). Knowledge of a species demands not only a definition of the mean or average structure and life-history of individuals, but also an estimate of the amount of variation which they show. Some species show little variation ; others a great deal.

Examples of Variation in Living and Fossil Species

In the starfish *Oreaster reticulatus* the number of arms may be four, five, six, or seven.

In the butterfly *Heliconius erato* the colour and pattern of fore and hind wings varies individually and racially.

In the snail *Cepaea nemoralis* some shells show dark bands while others lack them. In the banded shells the number of bands may be one, two, three, four, or five, or the bands may be confluent. In the plain shells the ground-colour may be cream or pale yellow or brown.

A parallel range of variation is found in the snail *Cepaea hortensis*.

Variation occurs in fossils as well as in living organisms, for instance in the sea-urchin *Micraster coranguinum* from the Cretaceous period (about 100 million years ago), or the Lamp-shell *Epithyris oxonica* from the Jurassic (150 million years ago). In the former, variation may affect the shape and structure of any part of the body ; in the latter the line of closure between the two valves of the shell may be straight or more or less undulated.

Causes of Variation

Variation may be caused by changes in the environment, or changes in the hereditary material, or both. Differences in environmental factors, i.e. outside the organism, such as climate, soil, humidity, food-supply, altitude, may produce changes known as *modifications* which are not inherited, and play no part in evolution. Differences in the hereditary material of the organism may produce genetic changes which are inherited and significant for evolution.

Variation Due to Climatic and Soil Factors

The extent to which characters show themselves in an individual plant often depends upon its conditions of growth, its environment. Climatic and soil conditions affect chiefly the size and form of growth and the reproductive capacity.

The effects of environment on growth can be studied by transplant experiments. By dividing a single plant into pieces it is possible to obtain a number of plants of the same age and genetical constitution, together forming what is called a *clone*. If members of the same clone are transplanted to different places, or are given different growing conditions at the same place, the differences arising between them are due entirely to the environment.

The effect of different altitudes and climates are shown in the exhibit on the right, the effects of different soils at the same place are shown on the left.

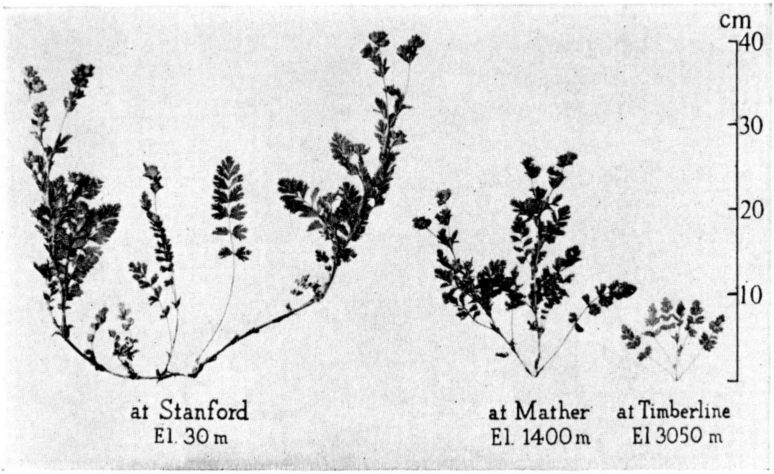

Figure 3. Variation due to Environmental Factors. The effects of transplanting *Horkelia californica* to different altitudes (expressed in metres).

Photograph by courtesy of the Carnegie Institution of Washington.

TRANSPLANT EXPERIMENTS AT DIFFERENT ALTITUDES

Numerous experiments made by transplanting members of the same clone to different altitudes in California show that conditions favourable for some species are not so for others, and that different ranges of tolerance with different optimum points form part of their genetical constitution. *Penstemon pachyphyllus*, which grows wild at 7560 feet, became markedly dwarfed when transplanted to sea-level, and it was prevented by the short growing-season from ripening seed when transplanted to very high altitudes.

On the other hand *Horkelia californica*, which grows wild at 700 feet, throve luxuriously at sea-level, but became dwarfed at higher altitudes and usually died in winter. (J. Clausen, D. D. Keck, and W. M. Hiesey.)

TRANSPLANT EXPERIMENTS ON DIFFERENT SOILS

Plants of *Plantago major*, all of the same age and all obtained by vegetative multiplication from the same individual plantain, were planted on the same day on different soils at the Potterne Biological Station, Wiltshire. The flat meagre growth of plants on sand, with their small and few leaves, contrasts with the luxuriant growth of plants on clay soils with more numerous and much larger ascending leaves. Plants transferred from one soil type to another quickly change their habit of growth. (E. M. Marsden-Jones and W. B. Turrill.)

Variation Due to Mechanical Factors

A common oyster of the Queensland coast *Ostrea glomerata* shows great irregularity in the shape of the shell according to the substratum on which the spat settled, and the space available for its growth. Many of these varieties were originally looked upon as distinct species before it was realized that their differences are due entirely to environmentally caused variation in a single species.

Variation Due to Food-Supply

Among the most important environmental factors are the quality and the quantity of the food-supply. Feeding experiments on pigs have shown that differences in size and shape may be caused by changing the food. If, in a pure strain of pigs where hereditary variation can be discounted, some are given restricted amounts and others unrestricted amounts of the same food for equal periods of time, differences are found in the development of the bones. In specimens of the same age at sixteen weeks, those which have had an unrestricted diet will have absolutely larger and relatively more massive bones. When they are full-grown, however, pigs fed on a restricted diet have longer and narrower skulls, a longer lumbar region of the vertebral column (the ‘ loin ’), and larger hip-girdles than those fed on an unrestricted diet although they take a longer time than the latter to reach a comparable stage of development. (J. Hammond and C. P. McMeekan.)

Facing case

Genetic Variation

Variation can be proved to be genetic by breeding experiments. As first discovered by Gregor Mendel, the transmission of characters from parents to offspring conforms to a definite pattern, known as Mendelian Inheritance (see p. 11).

Genes: Carriers of Inheritance

Hereditary resemblances and differences are controlled by particles known as genes which are located in the chromosomes of the cells. Normally when cells multiply (in growth or reproduction) they divide themselves into new cells, each with the identical gene-structure of the original.

Mutation: The Origin of Genetic Variation

Occasionally, a single gene is not exactly duplicated at cell-division, but changes suddenly. Such changes are called *mutations*, and the mutant genes go on copying themselves in their new condition. Any fresh mutation in a reproductive cell may therefore have considerable effects on succeeding generations of progeny produced from it. Mutations are the source of hereditary variation and the raw material of evolution (see p. 12).

Genes mutate because they are not completely stable chemical compounds. The nature of the effects which the mutations produce in the body of the organism has nothing whatever to do with the conditions of the environment; in other words, mutation is *not adaptive* and is not a response by the organism to its environment.

Calculations have been made of the frequency with which the genes mutate, and it has been found that in organisms as different from one another as a bacterium, a maize-plant, the fruit-fly, and man, it is of the same order of magnitude in them all, roughly one normal gene in half a million mutates at each place in the chromosomes in each generation. This frequency may not seem high, but because variance is conserved it is high enough to provide the possibilities of recombination and permutations giving rise to variation and evolution.

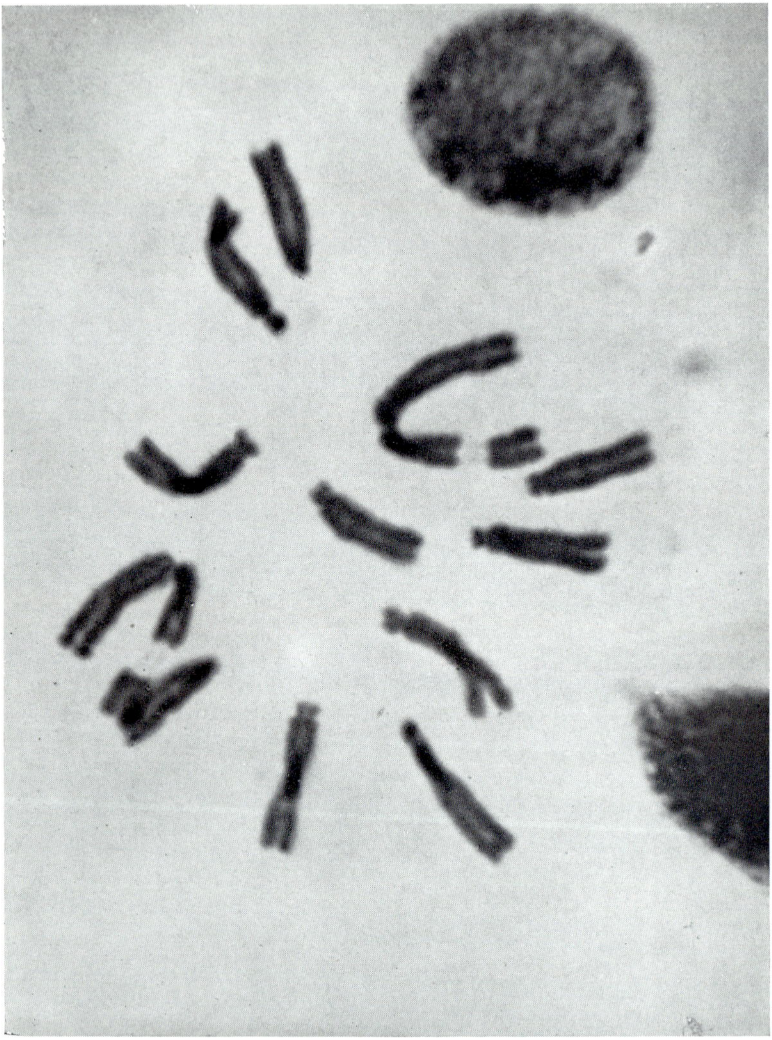

Figure 4. Chromosomes: Carriers of Genes. The chromosomes of the broad bean, *Vicia faba*, in process of division, magnified 2500 diameters.

Photograph by courtesy of Professor C. D. Darlington, F.R.S.

Recombination: Greatly Increased Variation

In all organisms produced by two parents, there are two corresponding sets of genes, one member of each pair having been derived from each parent. These pairs are separated (segregated) when germ-cells are produced, so that any egg or sperm contains only one member of each pair. When a sperm fertilizes an egg, pairs of genes are reconstituted. When the two members of a pair of genes are identical, the individual is called homozygous for that pair. As a result of mutation, however, the two members of a pair of genes may be different, in which case the individual is called heterozygous. At fertilization, therefore, there are many different possible recombinations of pairs of genes. This greatly increases the possible amount of hereditary variability.

Ten mutations in a species without recombination (without sexual reproduction) would result in ten novelties. Ten mutations in a species with the power of recombination would result in $2^{10} = 1024$ possible novelties.

The great advantages resulting from this power of recombination explain why species with separate sexes have evolved further (see p. 15).

Domestic Breeds

Differences between organisms may be due to breeding, as in the case of the Large White and Middle White pigs. The differences between such races are genetic and inherited, and must be distinguished from differences due to feeding which are not inherited, as shown in the neighbouring case.

Human Genetics

Human genetics is the study of the inheritance and effects of genes in man.

The following characters appear in children if they have inherited the relevant genes from even *one* parent. Such genes are called dominant.

The AB blood-groups
Atrophy of the auditory nerve (deafness)
Coloboma of the iris of the eye
Habsburg jaw
Night-blindness

Premature baldness in men
Supernumerary teeth
White forelock in men
Woolly hair in Europeans

The following characters appear in children only if they have inherited the relevant genes from *both* parents. Such genes are called recessive.

The O blood-group
Albinism (one form)
Blue or grey eyes
Deaf-mutism (one form)
Ear-lobes adherent
Hare-lip and cleft palate
Idiocy (amaurotic)
Inability to taste phenyl-thio-urea
Polydactyly (supernumerary fingers and toes)

Determination of Sex

In the higher animals, sex is determined at fertilization by the quantitative balance between the number of male-producing and female-producing genes that the fertilized egg contains. In many groups including man, the female-producing genes are mostly carried in a special chromosome called the X-chromosome of which the female has a pair. The male has only one X-chromosome, and one Y-chromosome which is unimportant for sex-determination. The male-producing genes are carried in the remainder of the chromosomes, and they are stronger than the female-producing genes contained in one X-chromosome, but weaker than those contained in two X-chromosomes. All eggs contain one X-chromosome ; half the sperms contain one X-chromosome and half contain one Y-chromosome. There are more or less equal chances that an egg will be fertilized by a sperm containing one X-chromosome and develop into a female, or by a sperm containing no X-chromosome but one Y-chromosome and develop into a male. The X-chromosome can often be distinguished in preparations of dividing cells, as in the photograph exhibited.

The Inheritance of Language-sounds

Some inherited characteristics are not demonstrable in single individuals, but may become apparent in whole populations, even requiring long periods of time to become detectable. Language-sounds belong to this category and may be illustrated by the distribution of the TH sound in Europe.

The map of TH distribution shows the areas in Europe where TH is pronounced, where it is variable, where TH was pronounced but was lost at the dates indicated, and where TH has never been pronounced during the past 2500 years.

The map of blood-group O distribution shows the areas in Europe where the frequency of the blood-group O gene in the population is higher than 65 %, between 65 % and 61 %, and lower than 61 %.

The close correlation between these maps shows that they both reflect the origins of the European peoples and that there may therefore be a genetic determination of the preference for certain sounds or, as it is said, a genetic component in the evolution of language. (C. D. Darlington.)

Left-hand case

How the Genes Work

Although the pairs of genes segregate independently of one another when they are reshuffled prior to the formation of a new generation, in the fertilized egg and the resultant plant or animal the genes exert their effects by interacting with one another and are affected by the conditions of the environment.

The nature of genes is explained in the case on the right. The mechanism by which genes are segregated and distributed to the offspring is exhibited in the fifth bay on the left in the Bird Gallery.

Multiple Effects of Genes

Genes can control more than one character. For instance, in the American Sulphur Butterfly *Colias philodice*, there is a gene which changes the yellow colour to white (in the female), and at the same time changes its behaviour by enabling it to fly actively at lower temperatures.

Conservation of Variance

The importance of Mendelian inheritance is that the genes are definite particles which collaborate with one another in controlling the characters of the organism, but never become contaminated (see p. 14). The offspring of a white and a red *Mirabilis jalapa* may be pink, but the offspring of two such pink flowers include whites and reds which are as pure as the original grandparents. It is this persistence of the genes which is responsible for the fact that when variation arises by mutation, instead of being continuously diluted by crossing until it disappears (as was once thought), it is conserved. (R. A. Fisher.)

Pure Lines

The proof that the effects of selection are due to different recombinations of heterozygous genes in the gene-complex is provided by the study of *pure lines.*

A pure line is a strain of plants or animals that is homozygous in all its genes. This condition can be produced by prolonged inbreeding, after which, selection produces no effects and the strain breeds true except for such mutations as may occur.

Bean seeds from the same pod of a pure line plant produce a big plant on rich soil and a small plant on poor soil. But bean seeds from such big or small plants when planted in the same soil produce identical plants.

The Gene-complex

Together all the genes in an organism form a balanced interacting system known as the gene-complex, each gene affecting and being affected by all the others. It is in this way that genes which at their first appearance produce intermediate effects have been made to *become dominant* or *recessive*, or have been suppressed in their visible effects. Even the effect of one and the same gene may have been modified in different directions by the gene-complex, as in the case of the currant moth.

The Evolution of Dominance

The currant moth, *Abraxas grossulariata*, typically has its pattern superimposed on a white background. There is a variety, *lutea* in which the background is suffused with yellow ; *lutea* differs from *grossulariata* by a single gene.

When *lutea* and *grossulariata* are crossed, the hybrids or heterozygotes are almost intermediate in colour, but they show a certain amount of variation which is due to the other genes in the gene-complex. If the white hybrid variants are selected for breeding, they produce a strain in which the heterozygote is almost indistinguishable from the *grossulariata* homozygote. In this strain the *grossulariata* gene has become dominant. Conversely, if the yellow hybrid heterozygotes are selected for breeding, they produce a strain in which the heterozygote is almost indistinguishable from the *lutea* homozygote. In this strain the *grossulariata* gene has become recessive and the *lutea* gene dominant (see p. 13). (E. B. Ford.)

The manner in which genes *gradually become* dominant or recessive as a result of selection of other genes in the gene-complex, shows how selection working on inherited variation can produce gradual change.

An example of such a change observed as having taken place in nature during the last hundred years is the Peppered Moth *Biston betularia*. The melanic form *carbonaria* first occurred in 1848, and the earliest preserved specimens are less intensely black than those found today, because they show traces of the light pattern of the typical Peppered Moth. The modern heterozygote is usually nearly all black. The evolutionary significance of this change is demonstrated in the next bay.

Reactions of Genes with the Environment

In addition to their interaction with the other genes of the gene-complex, genes are dependent on the factors of the environment in producing their various effects.

In the Himalayan rabbit there is a gene which is responsible for the production of a white coat, at the normal body-temperature of mammals. At lower temperatures this gene produces black hair. The Himalayan rabbit is born white all over since it has been developed at the temperature of its mother's womb. After birth, however, its extremities which are subject to chilling, ears, muzzle, and feet turn black.

A similar case is provided by the Siamese cat, where the tip of the tail is also subject to darkening.

East side: second bay
Right-hand case

Adaptation

All organisms are adapted to their environment. Man is adapted to life on land, a fish to life in water (see p. 9). In many instances, however, organisms possess structures which show a conspicuous physical relation to their environment.

Any habit or structure which enables its owner to survive and to leave the optimum number of offspring is an adaptation. Heritable adaptations have arisen through evolution in organisms which formerly lacked them.

Examples of Adaptation

THE WOODPECKER

The woodpecker is an example of adaptation to which attention was called by Darwin. It possesses four structures which enable it to carry out its feeding habits with greater efficiency : two of its toes point backwards, enabling it to get a foothold on the bark of trees ; its tail-feathers are stiff, enabling it to prop itself securely against the tree ; its beak is long and strong, enabling it to chisel holes in the bark ; its tongue is very long, enabling it to reach and catch insect-grubs at the bottom of the holes.

DESERT PLANTS

Plants inhabiting deserts show a reduction in leaves, thickening of the stems, and a smooth surface, all of which tend to minimize the loss of water by evaporation. This may occur in species of many different families, which although resembling each other in these adaptations, are not closely related to each other.

REGENERATION-JOINTS

In some animals with long appendages liable to be caught by predators, such as the limbs of crabs or the tails of lizards, special joints are formed where the appendages break, enabling the animals to escape. This is known as *autotomy*, and the joints are called autotomic joints. After its loss, the missing appendage is then formed anew by regeneration.

SEED-DISPERSAL

The dispersal of seeds is an important factor in the survival of species of plants, and use is made of numerous agents to ensure it, including wind, water, birds that eat the fruits and distribute the seeds with their droppings, and mammals that spread the fruits caught up in their fur.

FLIGHT

Insects, pterodactyls, birds, and bats are (or were) able to fly in the air by means of their wings. In each of these kinds of animals the wings are constructed on a different pattern.

PROTECTIVE RESEMBLANCE

Some animals show a remarkable degree of resemblance to the structures on which they normally rest, such as leaves, twigs, or bark, from which they derive protection owing to the inability of their predators to see them in their camouflaged condition. Examples are leaf-insects, stick-insects, caterpillars resembling sprouting twigs, bugs imitating thorns, and moths looking like the droppings of birds on leaves.

Facing case

Complex Adaptation

Some adaptations involve not only the relations of an organism to its environment, but require the collaboration of two or more organisms. These may be members of the same species, as in the male and female Huia-bird, or they may be a plant and an animal, as in the fig and the wasp.

Müllerian Mimicry in Butterflies

The individuals concerned all belong to species which are unpalatable, but by resembling each other they reduce the losses which each would otherwise suffer from the experiments as a result of which the predators learn that they are unpalatable. An attack on *one* of these unpalatable butterflies might discourage a predator from attacking any of them.

Delayed Moult in the Male Ptarmigan

Adaptation may result from simple alteration of the time-scale. In the ptarmigan, the moult from winter white to summer brown tends to be delayed in the male, with the result that he remains conspicuous while the female, biologically more valuable when incubating eggs, is inconspicuous. (J. S. Huxley.)

Mistletoe

Mistletoe is an example of adaptation involving plants and animals to which Darwin called attention. Mistletoe is a parasitic plant, entirely dependent on its host tree for supplies of water and salts from the soil. For its continuance, mistletoe depends upon three factors: an insect, usually a bee or a fly, for the pollination of its flowers; a bird, usually a mistle-thrush, for the dissemination of its seeds by eating the berries and depositing them on branches of trees ; a tree, usually a pine, a spruce, or a broad-leaf tree (according to the race of mistletoe), on which its parasitic relationship can be established.

Figure 5. Adaptation in the Huia-bird. The male bird on the left has a short and stout beak with which it chisels holes in trees, while the female on the right has a slender long beak which reaches the grubs at the bottom of the holes.

The Huia-bird

The Huia-bird of New Zealand is an example of adaptation involving both sexes but without reference to reproduction. The male bird has a short, stout beak with which it chisels holes in trees containing grubs of the particular beetle on which it feeds. The female has a slender curved beak, twice as long as that of the male and therefore with a longer reach, but not capable of chiselling. Co-operation between the members of a pair of mated Huia-birds is therefore indispensable for both to obtain food.

It is believed that the Huia-bird is now extinct; probably its extinction was due to its excessively specialized adaptation, preventing it from exploiting other possible food-supplies when the grubs on which it fed exclusively became less frequent as men destroyed or diminished the New Zealand forests (see p. 20).

The Pollination of Orchids

The function of insects in securing cross-pollination of flowers is well-known. Attracted to the flowers by their colour or smell and in search of nectar and pollen, insects insert their proboscis into the flower where they become dusted with pollen which is then conveyed to the stigma of the next flower visited by the insect.

In orchids an additional adaptation has been evolved. The pollen-grains are arranged in packets attached by elastic threads to a stalk, the whole forming a *pollinium* of which there are 2 to 8 in a flower. When an insect, usually a bee, a wasp or a moth, visits the flower and inserts its proboscis, the pollinia become detached from the flower and attached to the insect by a viscous fluid that rapidly sets hard. As the insect withdraws and flies away it carries the pollinia with it. The stalk of each pollinium bends down, and the pollinia then point forward on the insect and acquire a position where they come into contact with the stigma of the next flower visited.

Some orchids have flowers adapted to pollination by a single species of insect.

The Pollination of the Fig

The reproduction of the fig depends on the pollination of its female flowers with pollen from the male flowers of another tree. This is carried out by the fig-wasp *Blastophaga psenes* which develops parasitically in certain modified female flowers. As the female wasp leaves the fig in which she developed in order to lay eggs in another fig, she becomes dusted with pollen and carries it to the female flowers of the fig that she next visits.

Left-hand case

Polymorphism

Of some species more than one form occurs in the same area, and these forms may be adapted to different environments in which their adaptations are maintained by natural selection. Such species are said to show polymorphism.

Polymorphism may be of two kinds, transient or balanced. In transient polymorphism a genetically controlled condition is in process of spreading through the population and may ultimately lead on to uniformity. The industrial melanism of the Peppered Moth is an example of transient polymorphism.

In balanced polymorphism two or more forms are maintained in the same population by a balance of contending advantage and disadvantage. In such a condition the heterozygote is favoured as compared with either homozygote, and this maintains a high degree of diversity in the species. The blood-groups of man, sickle-cell, Batesian mimicry, and the land-snail (*Cepaea nemoralis*), are examples of balanced polymorphism.

Sickle-cell in Man

Amongst the natives of Central Africa, many have red blood-corpuscles which resemble curved sickles. When such people are out of breath, their red blood-corpuscles are destroyed, resulting in anaemia and, in serious cases, in thrombosis.

The sickling is due to a single Mendelian gene, which, when inherited, from one parent only (*sickle trait* . . . heterozygous) shows its effect only under extreme conditions of oxygen deficiency. However, when inherited from both parents (*pure sickle* . . homozygous) the effects of the gene are shown even under normal conditions.

Death from these causes removes a great many sickle-cell genes from the population in every generation. The reason why sickle-cells persist is because the heterozygous carriers of the gene are more resistant to infection by the Malaria parasite *Plasmodium falciparum*, and are therefore favoured by natural selection (see p. 24).

In an area of given malarial intensity, the proportion of sickle-cell people in a population reaches and maintains an equilibrium when the death-rate from sickle-cell anaemia exactly counterbalances the death-rate from malaria.

Death-rate from anaemia depends on the frequency of pure sickle. Malarial death-rate depends on malaria intensity and the percentage of naturally immune sickle-cell persons in the population. The highest equilibrium frequency of sickle-cell in West Africa is about 20 %, but among the negro population of the United States of America, where there is no endemic malaria, it has fallen to 9 %. (A. C. Allison.)

Industrial Melanism

Industrial melanism is the name given to the phenomenon in which moths are changing their complicated patterns from light to an all-black

Figure 6. Industrial Melanism. On the left is a tree-trunk covered with lichens from a rural area on which are one typical light-coloured Peppered Moth and one melanic form. On the right is a tree-trunk covered with soot from an industrial area on which are one typical light-coloured Peppered Moth and one melanic form.

Photograph of an exhibit in the Museum prepared by H. B. D. Kettlewell.

coloration. Of the 780 species of British Macrolepidoptera, more than 70 are in the process of doing this.

Industrial melanism is the most striking evolutionary change actually witnessed, and it demonstrates the effects of natural selection in producing adaptation conferring survival value, in accordance with Darwin's theory (see p. 22).

Tree-trunks covered with lichens occurred all over Great Britain before the Industrial Revolution. Today they are only found in un-

Figure 7. Selective Predation. A robin taking a melanic form of the Peppered Moth from a tree-trunk in Dorsetshire.

Photograph by courtesy of N. Tinbergen.

polluted areas such as the West of England and the Highlands of Scotland. The typical light-coloured Peppered Moth (*Biston betularia*) when resting on such a trunk or bough by day is almost invisible to birds, and thereby protected, whereas the melanic form *carbonaria* is extremely conspicuous and rapidly eliminated.

Since the Industrial Revolution, the atmosphere of many areas in Great Britain has become progressively polluted by smoke. In and around industrial areas the pollution is measured in tons per square mile per month. This has resulted in the disappearance of visible lichens from trunks and boughs of trees, and their darkening due to deposition of soot. The typical light-coloured Peppered Moth when resting on such trunks and boughs by day is extremely conspicuous and is rapidly eliminated by birds, whereas the melanic form *carbonaria* is protected on the dark background.

The first melanic *carbonaria* form of the Peppered Moth was taken in Manchester in 1848. By 1900 the proportions of the melanic *carbonaria*

Figure 8. Selective Predation. A redstart taking a typical light-coloured Peppered Moth from a tree-trunk near Birmingham.
Photograph by courtesy of N. Tinbergen.

form to the typical light-coloured Peppered Moth in Manchester was approximately 99 to 1. This represents a 30 % advantage of the melanic over the light-coloured form in Manchester for this period of fifty years. In addition to the advantage which the melanic *carbonaria* form derived from its protective coloration, it has also been shown to differ from the light-coloured form in its physiology and behaviour. Meanwhile, in unpolluted areas the typical light-coloured form was the only one found except for occasional melanic *carbonaria* mutants.

A map of the distribution of the varieties of the Peppered Moth shows that today there is a correlation between the industrial areas and a high proportion of the melanic *carbonaria* form. Furthermore, the *carbonaria* form never drops below 80 % of the population throughout the Eastern Counties of England. This is brought about by the indirect effects of smoke-drift from the industrial areas, due to the prevailing south-westerly wind.

Selective predation of the varieties of the Peppered Moth by birds has been directly observed, and may be illustrated by photographs of a robin taking a melanic *carbonaria* from a lichened tree-trunk in Dorsetshire on which there were two typical light-coloured forms which the bird missed, and of a redstart taking the typical light-coloured form on a heavily polluted tree-trunk near Birmingham on which there were three melanic *carbonaria* forms which the bird missed. The melanic *carbonaria* form enjoys a 10 % advantage in polluted areas, and suffers under a 17 % disadvantage in unpolluted areas. (H. B. D. Kettlewell.)

Batesian Mimicry in *Papilio dardanus*

Batesian mimicry is the resemblance of certain individuals of a species to individuals of another species whereby increased survival is obtained. The mimics belong to species which are subject to attacks by predatory birds and other animals, and they acquire relative immunity by imitating other species which are shunned by predators because of their unpalatability.

In most parts of its range, females of the species *Papilio dardanus* are polymorphic and present a number of forms, some of which mimic other species very closely. The different types of females are genetically controlled.

The proof that this type of mimicry is an adaptation that confers survival-value, and is subject to natural selection, is provided by the fact that in those parts of its range where, as at Entebbe, the relative frequency of models to mimics is high, the variability of the mimics is low, not exceeding 4 %; whereas in regions where the models are rare, as at Nairobi, where they are 70 times rarer than at Entebbe, the variability of the mimics is 8 times greater or 32 %. Where the model is rare and therefore relatively unknown to predators, the survival-value of resembling it is small, and there is less selection-pressure exerted towards accurate resemblance to the model (see p. 23).

The male does not mimic any other insect, and in Madagascar where the female is identical with the male, she is non-mimetic. (E. B. Ford.)

Primroses with Long and Short Styles

In Primroses, some plants have flowers with long styles and short stamens (' pin-eyed '), while others have short styles and long stamens (' thrum-eyed '). Darwin showed that this dimorphism, known as heterostyly, is an adaptation resulting in cross-fertilization between different individuals.

When an insect visits a short-styled flower, it is dusted with pollen on that part of its body which comes into contact with the stigma of long-styled flowers, and vice-versa. Cross-fertilization produces more vigorous and fertile offspring than result from self-fertilization. Heterostyly therefore encourages out-breeding.

East side: third bay
Right-hand case

Artificial Selection

Artificial selection is the deliberate breeding by man from certain variants of his domestic plants and animals rather than from others, with the object of emphasizing the characters which he wants. It has produced the difference between the cabbage and the cauliflower, the cart-horse and the Shetland pony, the Newfoundland dog and the Pekingese.

Observation of the effects of variation and selection in domestic plants and animals led Darwin to look for an analogous kind of selection in nature.

The Artificial Selection of Cereals

THE HOLDING QUALITY OF CEREALS

The basic distinction between cultivated cereals, such as wheat, barley, and rye, and their nearest related wild species is in the stalk of the ear which holds the ripened grains, i.e. its axis or rhachis. This is usually tough and retentive in the cultivated forms and fragile in the wild species.

In most wild grasses the axis shatters easily when ripe, scattering the

5

grains. From time to time they produce variants with a more retentive axis. This limitation of its means of dispersal is disadvantageous to an annual grass in a wild state ; under cultivation, however, forms with an axis which breaks up late or not at all are preferable, their ears being easier to harvest without loss of grain. Such variants are automatically better fitted for survival under human care than their fragile ancestors. A mutation of this kind made possible the perpetuation and development of the basic diploid wheat, Einkorn, as a crop.

WIND-POLLINATION AND HYBRIDIZATION

Cereals, in common with other grasses, have wind-pollinated flowers, their light dry pollen being air-carried from their well-exposed anthers to their equally well-exposed feathery stigmas. Pollen of different species can thus drift on to the stigmas with the result that a cereal crop may receive new characters by hybridization with related plants growing nearby. The diversity of the cultivated wheats is believed to be partly the result of such hybridization between species which are classified into the related genera *Triticum*, *Aegilops*, and *Agropyron*.

Artificial Selection in Wheat

Artificial selection by man has given direction to the evolution of the cultivated wheats along lines which are opposite to those which natural selection would have imposed in the wild state, with the result that man has obtained forms which are eminently suited to human needs but which cannot survive without his care.

The cultivated wheats provide the daily bread and other cereal foods of the greater part of humanity. Their cultivation began in Neolithic if not late Mesolithic times, long before the earliest civilizations. On account of this antiquity, their origin is obscure. They do not match any plant known in a wild state, but their wild ancestors were evidently grasses native to Western Asia. From these, by mutation, by natural hybridization, and by selection to meet human needs over thousands of years, and recently by deliberate controlled breeding, the wheats have evolved into many thousands of kinds (cultivars and strains).

WHERE WHEAT ORIGINATED

In Western Asia occur species of wild grasses of the Tribe Hordeeae which are believed to be ancestral to the cultivated wheats. These are

Triticum boeoticum (synonym *T. aegilopoides*), *T. dicoccoides*, *Aegilops squarrosa*, *A. speltoides* and possibly an *Agropyron* species.

Primitive people in various parts of the world gather seeds of wild grasses for food. The discovery by prehistoric men that a seed would produce a new plant, giving seed of the same kind when sown and tended, made civilization possible. To provide a start for a grain crop, the grass must be one which will thrive if accidentally dispersed around human dwellings, and then yield enough to encourage deliberate harvesting, storing, and sowing.

The first accidental cultivation of such grasses may have been near Mount Carmel, where have been found flint sickles used about 8000 years ago. Ancient Egyptian tradition relating to the god Osiris also indicated Syria or Palestine as the region where wheat or barley were found wild and first cultivated. In Syria grew up the cult of the Corn God Tammuz (Adonis).

CHROMOSOMES AND HYBRIDIZATION

The living material of a plant, like that of an animal, consists of microscopic cells each with a *nucleus* composed of smaller bodies, the *chromosomes*, which carry the factors for heritable characteristics and become evident when the cell divides to form new cells. The number and type of the chromosomes are constant for each individual species.

Ordinary cells have two sets of chromosomes, symbolized as $2n$. Individuals containing $2n$ chromosomes are called *diploid*, but the reproductive cells (pollen and egg-cell) normally contain only one set (n), which is called the *haploid* number. Fertilization brings together one set from the male and one from the female ; thus the fertilized egg-cell has a double set ($n+n$), and the facts can be tabulated as follows :

Ordinary cells (diploid)		*Pollen (haploid)*		*Fertilized egg (diploid)*		*Ordinary cells (diploid)*		*Pollen (haploid)*
$2n$	→	n			→	$2n$	→	n
		Egg-cell (haploid)	→	$2n$				*Egg-cell (haploid)*
$2n$	→	n			→	$2n$	→	n

The basic chromosome-set of one species may be represented by the letter A, of another by the letter B, etc. When, through hybridization, a chromosome-set (A) of one species is mated with a set (B) from another species, sterility may result unless, as happens occasionally though rarely, both sets are immediately doubled, a condition known as *polyploidy*, resulting in *tetraploid* individuals.

	Ordinary cells		*Pollen*		*Fertilized Egg-cell of Hybrid*		*Ordinary cells after doubling*
Diploid *Agropyron* or *Aegilops*	BB	→	B				
			Egg-cell	→	B A	→	BB AA
Diploid *Triticum*	AA	→	A				

	Ordinary cells		*Pollen*		*Fertilized egg-cell*		*Ordinary cells*
	AA BB	→	A B				
Tetraploid Wheat			*Egg-cell*	→	AA BB	→	AA BB
	AA BB	→	A B				

It will have been noticed that the term *-ploid* is used to indicate the number of basic chromosome-sets in ordinary cells of the plant-body ; a *diploid* has 2 sets, a *triploid* 3 sets, a *tetraploid* 4 sets, a *hexaploid* 6 sets, etc. A *polyploid* is a plant with more than 2 sets.

By their chromosome-numbers the wheats are classified into three main groups :

Diploid or Einkorn group ($2n = 14$)
Tetraploid or Emmer group ($2n = 28$)
Hexaploid or Spelt group ($2n = 42$).

It is believed that groups with high chromosome numbers are descended from groups with lower numbers.

DIPLOID OR EINKORN GROUP

The wild *Triticum boeoticum* and the cultivated Einkorn (*T. monococcum*) have 7 chromosomes (i.e. set A) in the reproductive cells, 14 (AA) in the ordinary cells.

The most primitive of wheats, closely related to the wild *Triticum boeoticum* of the Balkan Peninsula and Western Asia, Einkorn was widely grown in Neolithic times, but on account of its low yield it has been superseded except in barren upland regions and on poor soils where other wheats fail.

TETRAPLOID OR EMMER GROUP

The wild species *Triticum dicoccoides*, the cultivated Emmer (*T. dicoccon*), and Macaroni Wheat (*T. durum*) have 14 chromosomes (set AB) in the reproductive cells, 28 (AABB) in the ordinary cells. Of the 14 chromosomes, half (A) correspond to the 7 of the Einkorn group. The other half (B) are believed to have come from a weed grass, *Agropyron* or *Aegilops*, through hybridization of this with an Einkorn wheat, followed by chromosome-doubling.

Emmer is one of the oldest cultivated cereals and was apparently for thousands of years the only wheat grown in Ancient Egypt, where it is frequently depicted in tomb paintings. The carbonized wheat grains excavated from ancient tombs have lost all power of germination. The so-called ' mummy wheat ' offered to tourists as having kept its viability for thousands of years is really modern-grown wheat. Emmer is uncommon in modern Egypt. During Neolithic times, Emmer, in association with Barley and Einkorn, was widely cultivated in Europe and Western Asia. Although grown for human food in Abyssinia, India, and Southern Russia, and elsewhere here and there on a small scale, it is economically unimportant today. Emmer wheats have been derived from a wild species *T. dicoccoides* of Western Asia which in turn may have arisen from another wild species *T. boeoticum* by hybridization with an *Agropyron* or *Aegilops* species. In very early Neolithic times, wheats of intermediate character, very near the present wild *T. dicoccoides* and much coarser than the oldest Emmer known so far, were cultivated at Jarmo, Iraq. (H. Helback.)

HEXAPLOID OR SPELT GROUP

Spelt (*Triticum spelta*) and Bread Wheat (*T. aestivum*) have 21 chromosomes (set ABD) in the reproductive cells, 42 (AABBDD) in the ordinary cells. Of the 21 chromosomes, 7 (D) are believed to have come from a weed grass, *Aegilops*, through hybridization of this with a cultivated Emmer wheat, followed by chromosome-doubling whereby the sterile triploid became a fertile hexaploid.

Spelt is a wheat which apparently originated during the Bronze Age in Central Europe, north of the Alps. It was grown by the lake-side dwelling people of Switzerland and southern Germany, and was

	Ordinary cells		*Reproductive cells*		*Fertilized egg-cell of Hybrid*		*Ordinary cells after doubling*		
Tetraploid Wheat	AA BB	→	A B	} →	A B D	→	AA BB DD	=	Hexaploid Wheat
Diploid *Aegilops*	DD	→	D						

	Ordinary cells		*Pollen*		*Fertilized egg-cell*		
Hexaploid Wheat (*contd.*)	AA BB DD	→	A B D	}			
			Egg-cell	BB	AA BB DD	=	Bread Wheat Spelt
	AA BB DD	→	A B D	}			

apparently taken by Teutonic tribes to northern Spain. Although yielding less than Bread Wheat it is hardier, surviving frosts which destroy other wheats, and is more resistant to fungal diseases.

Bread Wheat is the most widely cultivated of cereals, surpassing all other wheats in the suitability of its flour for making light porous bread. To this species belong more than 3000 kinds (cultivars). They exhibit an enormous range of morphological diversity and of suitability to different climatic conditions. Vast acreages of Canada, Australia, and Russia are devoted to Bread Wheat.

In Bread Wheat the grain is loose within the spikelets of the ear and is thus easily separated from the chaff on threshing; it is technically a 'naked wheat', unlike the more primitive 'glume wheats' such as Spelt, Emmer, and Einkorn, in which the chaff (glumes) firmly encloses the grains.

Artificial Selection in Rye

Rye is believed to have originated over 2500 years ago in Asia Minor from *Secale ancestrale*, a species with a fragile axis to the ear but fairly large grains. It grows wild in sandy places, and plants of this species which invaded cultivated areas as weeds among crops of Wheat and Barley became subject to the same process of selection as the crops. Forms with a more retentive axis were harvested with the Wheat crop and threshed and sown with it. These weed ryes of Asia Minor formed the basis of cultivated Rye.

When the cultivation of Wheat extended northwards, Rye accompanied it as a weed, since the two cereals could hardly be separated by primitive agriculturists. The ability of Rye to thrive on poor soil and to resist frost and drought better than Wheat, Oats, and Barley, enabled it to yield a crop when these failed, particularly during the periods of climatic deterioration. In this condition, natural selection by climatic and soil factors eliminated the other cereals, leaving Rye as the sole crop. It was then no longer a weed.

Meanwhile, artificial selection by man favoured the forms with the greatest production of grain.

The evolution of Rye exemplifies the effect of natural and artificial selection upon mutant genes occurring in man-made conditions unaccompanied by either hybridization or change in chromosome-number. Lacking the causes of increased variability as compared with Wheat, the kinds of Rye number hardly more than fifty.

Rye (*Secale cereale*) is an important cereal crop in northern upland regions climatically too harsh for the profitable cultivation of Wheat. In Scandinavia and Central and Eastern Europe it provides flour for bread-making. In Britain it is grown primarily for cattle-feeding, but the straw furnishes good thatch, covers for bottles, etc. Even today northern farmers sometimes plant a mixture of Rye with other cereals, the relative success of one or the other depending on growing conditions during the year. In the Swiss Alps, Rye has been grown at altitudes up to 7000 feet.

Artificial Selection in Pigeons

The wild Rock-Pigeon *Columba livia* is the stock from which a large number of distinct breeds of the domestic pigeon have been obtained by man as a result of selection.

The variations which breeders have aimed to maintain, or whose features they sought to exaggerate, involve structure and behaviour. The Carrier and Racing-Homer have been selected for speed of flight and ability to return to their homes ; the Tumbler for ability to turn aerial somersaults ; the Pouter and Fantail for their remarkable courtship activities ; the Russian Trumpeter for its voice.

It is probable that the earlier domestic breeds came from the East, and some of the present breeds such as Barb, Fantail, and Tumbler can be traced back for three or four centuries.

In some cases this artificial selection by man has produced breeds that could not survive in a wild state. The Silver Owl breed has such a short beak that the parents are unable to feed their young which are fed by foster-parents. This condition is analogous to that found in Wheat shown in this case.

Facing case

Natural Selection

Natural selection is the effect of natural causes working on a population showing heritable variation, which results in the production of a greater number of offspring by that proportion of the variants that is better adapted to its environment. Although variation is random, selection channels it into adaptive directions, thereby producing the appearance of purposive change.

Natural Selection in the Land-Snail

Natural selection can be demonstrated in the snail *Cepaea nemoralis*, which is preyed upon by various animals hunting by sight, including thrushes. The snail is polymorphic, having many distinct varieties differing in the colour of the shell and the number of dark bands on it. Thrushes readily find the more conspicuous varieties and bring them to anvil-stones to break their shells. The proportion of the different varieties in any colony varies with the background, conspicuous varieties being kept rare by predators.

Against brown leaf litter, yellow snails are the most easily seen and captured. Yellow shells are less conspicuous in grass, where they are more common than dark shells. In coarse vegetation, striped shells are the least conspicuous and the most common.

Thrushes' anvil stones offer a good opportunity of seeing natural selection in action. By scattering snails on which inconspicuous marks have been made, the exact area where a thrush has been working can be determined, and the proportion it takes of each variety can be compared with those of the snail population of the area of the particular background where they were taken. In a woodland colony it has been shown in this way that the yellow colour of the shells actually changed from being disadvantageous to advantageous as the background changed from brown to green as the season of spring advanced. This example shows that the effects of selection may vary from place to place and from season to season (see p. 22). (A. J. Cain and P. M. Sheppard.)

Left-hand case

Sexual Selection

The theory of *Sexual Selection* was put forward by Darwin to explain the origin of striking colours, structures, behaviour, etc., found only in one sex of some animals (usually the males), and used in fighting for mates and in the activities of courtship. According to this theory, individuals with such characters well developed benefit from greater breeding success and more offspring. In the majority of cases, however, these characters are not the result of sexual selection in Darwin's sense, so much as of natural selection favouring general reproductive activity through conspicuous recognition marks, virile behaviour, etc.

Examples of Sexual Selection Benefiting One Sex

The Ruff is a sandpiper in which the males have frills of conspicuous feathers which they can erect. At the breeding season the males stake out their claims of territory, fight male intruders, and court females with elaborate display. The more active and strikingly coloured males are the more successful in acquiring mates.

The male Lyre-bird has a tail made of remarkably specialized feathers. It builds a mound of earth on which it stands and sings, and displays to the female by spreading its tail forward over its head and making the filaments of its feathers vibrate so fast that they look like a shimmering haze of light.

The male Amherst Pheasant makes a spectacular courtship display in front of females and arouses their interest in him for mating. His display takes special advantage of those plumage areas which are brightly patterned, and by which he is distinguished from the female. A similar display before other males serves as a threat towards them, and helps to eliminate his competitors.

Feathers from the male Argus Pheasant show varying degrees of pattern in different feathers from the same bird. They form a gradual series from small simple markings to large and spectacular eye-spots. This indicates the probable transitions through which the more spectacular feathers may have evolved in response to sexual selection.

Sexual selection does not depend on conscious choice by the female, but on whether the colour and display of the male has or has not succeeded in stimulating her to undergo mating.

Male Elephant Seals fight for breeding territories and for possession of the females which are smaller than the males. Size seems to be a considerable advantage to males of this species in their violent competitions on the beach, and larger males seem to be more successful than smaller individuals. Fighting for territory is usual amongst males of polygamous species.

The antlers of male Red Deer are shed and regrown every year, which causes a heavy drain on the resources of the body. There must therefore be some good reason why they have evolved. They are used for fighting other males, and as deer are polygamous it is deduced that their possession is conducive to reproductive advantage.

Examples of Sexual Selection Benefiting the Species

In the Great Crested Grebe, courtship activities with display frequently take place after as well as before mating. The sexes are similar in appearance and the display is mutual. One of the most striking examples of this is the ritual whereby both birds of a mated pair dive beneath the surface of the water, collect nesting material in the form of water-weed, bringing it to the surface in their mouths, swim at each other, meet breast to breast and rise up in the water with their bills touching and their ruffs fully spread.

Display of this kind keeps the couple together to share the raising of the young, and confers advantage on the species. (J. S. Huxley.)

In the spring, the male stickleback develops a nuptial dress of bluish white on the back and brilliant red on the throat and belly. It stakes out a territory, builds a nest, fights all other males, and courts and entices females to the nest to spawn. Turning on its side, the male points its head towards the entrance of the nest.

The red colour of the male acts as a threat to other males and an attraction to females ; the signal for the male to lead to the nest is given by the female when, heavy with eggs, she points her head upwards. (N. Tinbergen.)

Figure 9. Sexual Selection. The mutual display and ritual of the Great Crested Grebe.

Photograph by courtesy of Jac. Drenth, The Hague.

East side: fourth bay
Right-hand case

Continuity of Life

Darwin was greatly struck by the presence in South America of fossil animals with armour similar to that of the armadillo now living in that Continent (see p. 4). This Continuity of Type between fossil and living forms in the same area is explained by evolution, and it demonstrates the continuity of life throughout geological time.

Examples of Fossils and their Related Living Forms

The fossil brachiopod *Lingula* from the Ordovician of Wales, 400 million years old, may be compared with a specimen of a living species of the same genus. *Lingula* is one of the oldest genera known.

The fossil mollusc *Tryblidium* with the segmentally repeated muscle-scars on its shell, from the Upper Silurian of Sweden, 330 million years ago, may be compared with the living *Neopilina* which has segmentally repeated gills.

The fossil King-crab *Mesolimulus* from the Upper Jurassic Lithographic Stone of Bravaria, about 150 million years old, can be compared with a living *Limulus*. The King-crabs have persisted from Lower Devonian times to the present day, about 320 million years.

The living Australian Lung-fish *Epiceratodus* may be compared with a fossil *Ceratodus* about 180 million years old from the Triassic strata of New South Wales, Australia. Lung-fishes have persisted from the Devonian to the present day, about 320 million years, and are now found in the rivers and swamps of Australia, Africa, and South America.

The Coelacanth fish *Undina* from the Upper Jurassic Lithographic Stone of Bavaria can be compared with the living *Latimeria*. Coelacanth fishes have survived almost unchanged since the Devonian period, some 300 million years ago.

The fossil Maidenhair tree *Ginkgo* from the Middle Jurassic of Yorkshire may be compared with the living species of the same genus. The ginkgos have persisted from the Jurassic period, about 160 million years ago, and have a single living representative, the maidenhair tree of China and Japan.

The fossil *Glyptodon* from the Pleistocene beds of South America can be compared with the present day armadillo.

The wombat *Phascolomys mitchelli* was known as a fossil before living animals of the species were discovered in Australia.

Animals which have become Extinct since the Advent of Man

The giant sloth *Mylodon* from South America was a contemporary of man. Specimens still show fleshy muscle-attachments at the back of the skull, and portions of the skin have hair.

Animals on the Verge of Extinction

Thylacinus cynocephalus, a carnivorous marsupial (Tasmanian 'wolf'), is so near extinction that it has been seen alive only on rare occasions. It occurs as a fossil on the mainland of Australia where it no longer survives.

Facing case

Lines of Evolution

Research in palaeontology has revealed sequences of fossils that indicate the lines along which different groups of animals have evolved. The speed of evolution or rate at which forms have succeeded one another can be estimated from ages of the rocks as determined by measurement of radio-activity and other methods.

In general, an evolutionary line has four phases : *divergence* from other lines, *progress* in differentiation, *stability*, and *extinction.*

The Evolution of Ammonites

Ammonites are an extinct group of shellfish, related to modern squids, octopus, and nautilus, and usually having a spirally coiled external shell. Their evolution can be illustrated by that of a line represented successively by two families, the Liparoceratidae and Amaltheidae, which lived in the seas of the Lower Jurassic Period (Lias) about 165 million years ago. The time-span of the two families is three to four million years. The succession of forms naturally shows the order of their appearance in the Liassic rocks. The sequence shows not only evolutionary change but divergence into different evolutionary lines (see p. 24).

Liparoceras cheltiense is the root stock of the whole evolutionary sequence; its whorls are large and thick, and the inner whorls are almost entirely overlapped by the outer whorls. This type of ammonite persisted with little change up to the middle of zone C (*L. nautiliforme*).

Towards the top of zone A the two new species *L. contractum* and *L. naptonense* were evolved, both with more slender whorls and less overlap than in *L. cheltiense*, *L. contractum* being fine-ribbed and *L. naptonense* coarse-ribbed. This novelty affects the earlier-formed inner whorls.

In *Androgynoceras lataecosta* the whole ammonite takes the form of the inner whorls of *Liparoceras contractum*. The whorls hardly overlap each other at all, so that the inner whorls are completely exposed, and the ribs have become much stronger.

The genus *Oistoceras* differs from *Androgynoceras* solely in the forward reflection of the ribs on the back of the whorl, so that they form chevrons.

In the evolution of *Amaltheus*, the whorls became laterally compressed and the chevrons bifurcated. *Amaltheus bifurcus* is an intermediate form, with inner whorls shaped like those of *Oistoceras* and outer whorls more compressed as in *Amaltheus*. In *A. stokesi* the compression spreads to the inner whorls as well.

From *Amaltheus stokesi* three different lines of evolution can be traced: the *Amauroceras*, *Pseudoamaltheus*, and *Pleuroceras* lines.

In the descendants of *A. stokesi* the connexions between the ribs and their chevrons become less distinct. The *A. subnodosus*—*A. gibbosus* line leading to *Pleuroceras* consists of small, tuberculate forms, and the *A. margaritatus* line consists of large, highly compressed types.

Amauroceras is a degenerate line in which all the ornament is lost. It commenced when the smoother types amongst the variable young of *Amaltheus stokesi* retained their youthful characters and small size in the adult, forming the new species *Amaltheus wertheri*. Complete loss of the ornament in the latter then gave rise to *Amauroceras ferrugineum*, and at a higher level (middle of zone D) a much larger form appeared (*A. lenticulare*).

Pseudoamaltheus evolved from *Amaltheus margaritatus* by loss of the radial ribs and the chevrons, and by development of strong spiral ribbing.

The genus *Pleuroceras* differs from *Amaltheus* in having a quadrilateral whorl cross-section. The keel and its chevrons remain. *Amaltheus gibbosus* is the ancestor. *Pleuroceras transiens* is intermediate, and *P. solare* is the first true *Pleuroceras*. *P. solare* then evolved further into :

1. The *P. hawskerense* line, with slender whorls, and
2. The *P. spinatum* line, with massive whorls, later acquiring large tubercules on the inner whorls of *P. yeovilense*.

At the level of the Upper Cretaceous the group of ammonites became extinct. The reasons are not fully known, but in Britain extinction coincided with a sudden change in the types of sediment being laid down in all areas. It seems likely that the ammonites then existing were not able to evolve sufficiently rapidly to keep adapted to their changing environment and so to survive. (L. S. Spath.)

The Evolution of Horses

The evolutionary history of the horse can be traced step by step, showing its descent from a small browsing animal no bigger than a fox-terrier (see pp. 7 and 24).

The earliest horses lived in the Eocene period about 70 million years ago, and were about the size of a fox-terrier, with a shoulder height of eleven inches. *Hyracotherium* (or *Eohippus*) had four functional toes on each forefoot and three on the hind feet, probably adapted for life on soft marshy land. The molar teeth were low-crowned and suited only to browsing on soft herbage.

In the Oligocene period, about 45 million years ago, horses had increased in size and had a shoulder height of two feet. *Miohippus* had three toes on each foot with the central toe slightly larger than the other two. The low-crowned teeth and three-toed feet suggest they were still forms that lived in forests browsing on soft vegetation.

Horses of the Pliocene period, about 10 million years ago, had increased further in size and were about four feet high at the shoulder. They had high-crowned teeth closely resembling those of the modern horse *Equus*, and suited to grazing. The feet had a single large toe forming a hoof on each foot, and were adapted for running on hard ground. The side toes were still further reduced.

The genus *Equus* first appeared about a million years ago at the beginning of the Pleistocene period in North America, whence it spread rapidly to every continent except Australia. These animals became extinct in North America where they were re-introduced by man from Europe. The different species can be distinguished as true horses, zebras, and asses. (G. G. Simpson.)

Left-hand case

The Geological Succession

Fossils found in successive layers of the earth's crust show that the plant and animal life of the world has continuously changed. There has been a succession of floras and faunas, and the changes which the related groups of plants and animals have undergone are due to evolution (see p. 6).

The Cambrian is the earliest geological period from which well preserved fossils survive; its rocks contain seaweeds, worms, shells, trilobites and graptolites, but no land plants, insects or vertebrates.

Fish-like vertebrates first appeared in the Ordovician, true fish and land plants in the Silurian, amphibians and insects in the Devonian, reptiles in the Carboniferous, mammals in the Trias, birds in the Jurassic, flowering plants in the Cretaceous, and man in the Pleistocene period. Conversely, graptolites became extinct in the Carboniferous, trilobites in the Permian, pteridosperms in the Trias, dinosaurs in the Cretaceous.

The strata of the later geological periods do not contain many plants and animals of earlier periods. New forms were continually appearing while other forms became extinct.

A diagram of groups of plants and animals with restricted range in geological time shows the change in *emphasis* in the floras and faunas at different periods.

Examples of Time-Spans of Groups of Plants

Collenia, nodules of thin layers of carbonate of lime formed possibly by algae, from the Pre-Cambrian (Algonkian) of Montana, U.S.A. Undoubted algae are present in the Pre-Cambrian (Huronian) of Ontario, Canada, 1700 million years old. (S. A. Tyler and E. S. Barghoorn.)

Solenopora, an alga from the Jurassic of Morocco. Algae are among the earliest forms of life and have persisted from Pre-Cambrian times to the present day.

Neuropteris, a seed-fern from the Coal Measures of Radstock, Somerset. Pteridosperms persisted for 90 million years from the Upper Devonian to the Lower Triassic period.

Pterophyllum, a Cycad from the Triassic (Keuper) of Stuttgart, Germany. Cycads have persisted for 170 million years from the Upper Triassic to the present day, but were especially abundant during the Jurassic and Cretaceous periods.

Corylus, the hazel tree, a flowering plant from the Eocene of Mull, Scotland. Flowering plants and deciduous trees first appeared in the Cretaceous period, 100 million years ago.

Examples of Time-Spans of Groups of Animals

Calymene, a Trilobite from the Middle Silurian of Worcestershire. Trilobites are among the earliest fossils known and ranged from the Lower Cambrian, 500 million years ago, to the Lower Permian period, 200 million years ago.

Didymograptus, a Graptolite from the Ordovician of Pembrokeshire, Wales. The dendroid or net-like graptolites ranged from Upper Cambrian to Lower Carboniferous, but the true graptolites only occurred in the Ordovician and Silurian periods.

Stephanoceras, an Ammonite from the Middle Jurassic of Dorset. Ammonites first occurred in the Triassic period, abounded in the Jurassic, and died out in the Cretaceous.

Pteraspis, an Ostracoderm from the Lower Old Sandstone of Podolia, U.S.S.R. Ostracoderms, jawless fish-like animals, are found in estuarine or fresh-water deposits from the Middle Ordovician to the end of the Devonian period.

Amblypterus, a chrondostean fish from the Lower Permian of Czechoslovakia. Chondrostean or ray-finned fishes occurred in the Palaeozoic Mesozoic eras and there are a few present-day survivors such as the sturgeon.

Mene, a teleostean fish from the Upper Eocene of Monte Bolca, Italy. The teleostean or true bony fishes have persisted since the Jurassic period and include the vast majority of fishes alive at the present day.

Capitosaurus, a labyrinthodont amphibian from the Lower Trias (Bunter Stone) of Baden, Germany. Labyrinthodont amphibia ranged from the Upper Devonian to the top of the Triassic period.

Plateosaurus, a saurischian dinosaur from the Trias of Germany.

This group of dinosaurs persisted from the Triassic to the Cretaceous period.

Iguanodon, an ornithischian dinosaur from the Lower Cretaceous Weald Clay of Surrey. These dinosaurs persisted from the Upper Triassic to the Lower Cretaceous period.

Ichthyosaurus, a marine reptile from the Jurassic Upper Lias of Wurttemberg. Ichthyosaurs persisted from the Triassic to the Cretaceous period but were especially numerous in Jurassic times.

Plesiosaurus, a marine reptile from the Jurassic Lower Lias of Somerset. Plesiosaurs appeared in the Triassic and persisted through Jurassic to Cretaceous times.

Rhamphorhynchus, a pterosaur from the Upper Jurassic Lithographic Stone of Bavaria. Pterosaurs ('pterodactyls' or 'flying reptiles') lived in the Jurassic and Cretaceous periods.

Archaeopteryx, a bird from the Upper Jurassic Lithographic Stone of Bavaria. This, the earliest bird known, lived 150 million years ago and was a perfect mosaic of reptilian characters (tail, brain) and avian characters (feathers, foot). (See p. 25.)

The earliest mammals are found in the lowest Jurassic strata of the Rhaetic beds, 170 million years old.

Man emerged in the Pleistocene period, which began a million years ago.

East side: fifth bay
Right-hand case

Constancy of Plan

When vertebrate animals are compared, it is found that, however much they differ in size, shape, and manner of life, they are built on the same general plan and their parts correspond in relative position. This constancy of morphological relations in spite of differences in detail is due to the affinity between the animals through their descent with divergence and modification from a common ancestor.

Morphology

In the forelimb of vertebrates there is a general pattern of bones. This pattern consists typically of one bone in the upper arm (*humerus*), two in the forearm (*radius* and *ulna*), several in the wrist (*carpals*), five in the hand (*metacarpals*) to each of which are attached a few finger-joints (*phalanges*). This is true regardless of whether the animal is a frog, reptile, bird, or mammal, and lives in the water, on earth, or in the air (see p. 5).

Vestigial Organs

In many plants and animals there are vestigial organs, reduced in size and incapable of carrying out their original functions. The skeleton of the limbs of a snake is an example which shows that the snakes are descended from normal reptiles with functional limbs (see p. 5).

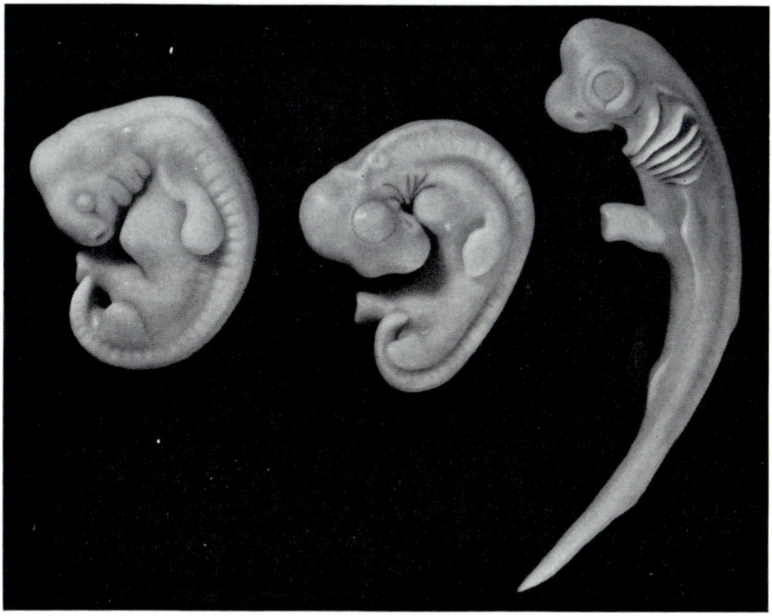

Figure 10. Embryonic Resemblance. Embryo of man on the left, chick in the centre, and dog-fish on the right, showing their similarity.

Embryology

Embryos of animals in the same group often resemble one another very closely even if their adults are very different in appearance. This resemblance is evidence of affinity between the animals and of their descent with divergence and modification from a common ancestor.

In the embryos of a dog-fish, a chick, and a man, the general plan of structure and the shape and position of the brain, eye, nostril, mouth, gill-slits or gill-pouches, forelimb, hindlimb, muscles, and tail, show great similarity whereas their adults are very different (see p. 5).

Paedomorphosis

It was at one time thought that the embryo of the descendant represented the adult of the ancestor, but this theory of ' recapitulation ' is now known to be false. Embryonic stages of descendants are generally the repetition of corresponding embryonic stages of the ancestors.

There are, however, many cases where the adult descendant resembles the *young* ancestor. This type of descent, called *paedomorphosis*, or ' Peter Pan ' evolution because the youthful characters of the ancestors are retained and retarded in the descendant which does not ' grow up ', is believed to have occurred in insects, vertebrates, and man (see p. 26).

The series of skulls of chimpanzee, australopithecine, *Pithecanthropus*, and Neanderthal Man, shows that in the young form of each, the brain-case is spheroidal, the forehead is smooth, and the snout small, whereas in the adult the brow-ridges are massive and the snout is prominent. Modern man when adult retains the youthful condition.

Other features in modern man showing the effects of paedormorphosis are the formation of the chin, the delay in eruption of the teeth, and the delay in fusion of the bones of the skull which enables the brain to grow.

Facing case

Geographical Variation

The study of species over wide geographical areas frequently reveals the existence of continuous gradients of change in the characters of organisms, observable over different parts of their ranges, and relating to colour, size, or other features. Such gradients of characters are called *clines*, a term introduced by Sir Julian Huxley, and include varieties such as *colour-clines* and *size-clines*.

Colour-clines

The deer-mouse *Peromyscus maniculatus* provides an example of a colour-cline. Along a west-east axis through the United States, the colour of these animals grades from dark brown on the West coast, through grey in the Eastern Rockies, becoming paler and more orange in the Prairies, and then greyer again in the Eastern Prairies and finally darker and browner in the Eastern Forests. This character-gradient is related to humidity and other climatic features.

Ring-Species

On the shores of the Arctic ocean in Eastern Siberia lives the Vega gull (*Larus argentatus vegae*) with dull yellow flesh-coloured legs and a dark mantle. Passing eastwards across the Bering Straits to North America, the Vega gull grades into the American herring-gull (*L. a. smithsonianus*) and across the Atlantic into the British herring-gull (*L. a. argentatus*) with flesh-coloured legs and a very light mantle.

On the other hand, in a westerly direction, the Vega gull grades into the Lesser black-backed gull (*Larus fuscus fuscus*) and its British race (*L. f. graellsii*). Around the chain the gulls interbreed, but at the ends of the chain in Great Britain, the lesser-black-backed gull and the herring-gull behave as separate species, differing not only in appearance but in behaviour (see p. 28). The former tends to breed inland on moors and to be migratory in winter, while the latter prefers to nest on cliffs and is dispersive in winter. (R. Meinertzhagen ; J. Fisher and R. M. Lockley.)

Size-clines

The Puffin provides an example of a size-cline ; the maximum size which those birds reach increases with the degree of Northerliness of their habitat, at a rate of 1 % of linear measurement per 2° North Latitude. This is illustrated in the length of the wings.

This size-cline is an example of Bergmann's Rule according to which warm-blooded animals increase in size in colder climates. This is an adaptation to diminution of heat-loss, since the larger the animal, the smaller is the ratio between its volume (which produces heat) and its surface (which loses heat).

Left-hand case

Isolation

Evolution takes place when, for various reasons, populations cease to breed with neighbouring populations, and, under different conditions of selection, accumulate heritable variations in different directions.

Biological isolation is a necessary condition for divergence leading to the formation of new species and higher groups (see p. 28).

Isolation may be *geographical* when physical features like oceans or mountain-ranges separate whole populations ; *ecological* when different modes of life such as feeding, breeding, and nesting habits, or adaptations to different temperatures or degrees of salinity separate portions of populations ; or *genetic* when sterility between groups of individuals results from sudden changes in the gene and chromosome mechanisms, due to mutation or polyploidy.

Geographical Isolation

THE FINCHES OF THE GALAPAGOS ISLANDS

Darwin visited the Galapagos Islands in 1835 (see pp. 4 and 28) and was immediately struck by two facts : the finches all showed a general resemblance to those found on the mainland of South America ; and some of the islands each had their own peculiar species of finch. There is no rational explanation of the similarity of the Galapagos birds to those of the mainland, nor of the difference between the birds on the different islands, except on the view that they are descended from a common stock, and, under conditions of isolation on the different islands, have become modified by local factors of selection. One species, *Geospiza pallida*, on Albemarle Island is a ground finch chiefly feeding on seeds ; *Camarhynchus habelli* on Abingdon Island is a tree finch feeding on insects. These observations led Darwin to suspect that species were not immutable but had arisen by evolution. (C. Darwin; D. Lack.)

THE RHINOCEROSES OF THE MALAYAN REGION

Darwin wrote in 1844 : ' Shall we then allow that the three distinct species of rhinoceros which separately inhabit Java and Sumatra and the neighbouring mainland of Malacca were created, male and female, out of the inorganic materials of those countries ? Without any adequate cause, as far as our reason serves, shall we say that they were merely, from living near each other, created very like each other . . . or have they descended, like our domestic races, from the same parent stock ? '

The animals to which Darwin referred are now known as *Rhinoceros sondaicus*, *Didermoceros sumatrensis sumatrensis*, and *Didermoceros sumatrensis lasiotis*.

Ecological Isolation

In the butterfly *Colias philodice* there is a form of female distinguished by being white instead of yellow, and physiologically adjusted to activity at a lower temperature than its normal counterpart. For this reason the white form is more frequently found in the early morning and in the evening. These differences are controlled by a simple genetic mechanism.

The yellow and white forms live in the same areas, and the partial separation between them is based neither on geographical nor on sterility barriers, but on genetic control adjusted to different environmental conditions, in this instance that of temperature, leading to the production of distinct forms which may perhaps evolve independently. (W. Hovanitz.)

Genetic Isolation

Even related plants living side by side may be isolated from each other by their inability to breed and produce fertile offspring. This inability may be caused by incompatibility between the genes and chromosome mechanisms of the two plants. This is usually the case with different species. Examples are shown in which a hybrid is isolated from both its parent species by a genetic process called *Polyploidy* (see p. 56).

The Origin of Some Species

Primula floribunda and *P. verticillata* are different species of primrose which can interbreed, but the resulting hybrid is usually sterile because the chromosomes derived from one of the parents do not match those derived from the other in the ' dance ' which they undergo preparatory to the production of reproductive cells. Occasionally, however, the chromosomes of the hybrid become doubled as a result of their division without a corresponding division of the cells containing them, and plants in which this has happened are called *allopolyploid.* In such hybrids the chromosomes all have compatible partners and fertile reproductive cells are formed which produce fertile offspring (see p. 55). *Primula kewensis* was produced in this manner from *Primula floribunda* and *P. verticillata* ; it breeds true, differs structurally from both *Primula floribunda* and *P. verticillata*, and is sterile with each of these species because of the genetic isolation resulting from chromosome incompatibility. *Primula kewensis* is therefore a species the origin of which has actually been observed (see p. 29).

Galeopsis tetrahit is a wild species with 32 chromosomes, intermediate between *G. pubescens* and *G. speciosa* each of which has 16. In a laboratory, crosses involving *pubescens* and *speciosa* were made and they produced a polyploid hybrid with 32 chromosomes which was not only similar to the wild *tetrahit*, but fertile with it. In this case a species which occurs in nature has been synthesized. (A. Müntzing.)

Other examples of plants in which new species have resulted from hybridization followed by chromosome duplication are the garden

Figure 11. The Origin of a Species. On the left a plant of *Primula floribunda*, in the centre a plant of *Primula verticillata*, on the right a plant of *Primula kewensis* obtained as a hybrid of the other two which has undergone doubling of its chromosomes.

Dahlia, the European plum *Prunus domestica*, the pink-flowered chestnut *Aesculus carnea*, the loganberry *Rubus loganobaccus*, the swede *Brassica napus*, and Townsend's cord-grass *Spartina townsendii*.

The formation of new species in animals by means of polyploidy must be rare. It has been claimed that there may be a case in the golden hamster, *Mesocricetus auratus*, which has twice the number of chromosomes of both the European hamster *Cricetus cricetus* and the Chinese hamster *Cricetulus griseus*, from which it was possibly produced in the Caucasian region where their ranges overlap. (L. Sachs.)

West side: first bay
Left-hand case

Classification

Plants and animals can be classed in groups which are contained within larger groups. This classification is not fortuitous like that of pebbles on a beach, nor arbitrary like the grouping of stars in imaginary constellations. The members of each group of plants and animals are placed together because they resemble each other, and the groups in turn are characterized by resemblance. The resemblance between organisms, or between groups of organisms, is due to their relationship (genetic affinity) ; the difference between them is due to descent with divergence and modification from ancestors shared in common. A natural classification, such as this, provides a family tree of relationship and reflects the course of evolution (see p. 6).

In this exhibit, a simple classification is given of the Chordata which include the vertebrates. The lower Chordata and Fishes are illustrated in greater detail with examples of the main groups and explanations of the structures by which they are classified.

The Chordata

The Chordata are grouped together because they (and no other animals) all have the following three structures :

an internal skeletal rod called a *notochord*,
a tubular nervous system running down the back, and
a series of gill-pouches on each side of the throat.

The Chordata may be divided into those animals without skulls which are called the Acrania, and those with skulls which are called the Craniata. The former include the Cephalochordata (*Branchiostoma*), Hemichordata (*Balanoglossus*), and Urochordata (Tunicates). The latter can be divided in those without jaws, the Agnatha, and those with jaws, the Gnathostomata. The former include the Cyclostomata (lampreys) and some extinct groups. The latter can be divided into those with fins, the Fishes or Pisces, and those with hands and feet ending in fingers and toes, the Tetrapoda.

As an example of a well-defined group, the Pisces may be taken and classified into their chief subdivisions, as follows.

Pisces (Fishes)

PLACODERMI, a diverse assemblage of extinct fishes, representing a number of early experiments in fish structure. Among them some had jaws in series with the gill-supports much like primitive sharks, but in others they seem to have been simpler. Some of them had paired fins, but these might be in any number from one to seven, and had not developed any of the structural characters which were to make fish fins so efficient either for balance or for locomotion.

CHONDRICHTHYES, fishes with the skeleton made of cartilage (gristle).

SELACHII, sharks and rays.
HOLOCEPHALI, rabbit fishes and their extinct ancestors.

West side: first bay
Right-hand case

OSTEICHTHYES, fishes with the skeleton largely of bone.

CROSSOPTERYGII, bony fishes in which the paired fins have a central axis, and primitively there are two dorsal fins. The heavy scales of primitive forms were of the kind known as cosmoid. In some, the nostrils communicate with the mouth.

RHIPIDISTIA, mainly piscivorous, among which some had limbs adaptable to crawling, lungs and nostrils adaptable to air-breathing, and which gave rise to the land-vertebrates or Tetrapoda. They also gave rise to:
COELACANTHINI, the Coelacanth fishes.
DIPNOI, which specialized in a crushing dentition and survive in Australia, Africa, and South America as the lung fishes.

ACTINOPTERYGII, bony fishes in which the paired fins are fan-shaped; and primitively there is only one dorsal fin. The heavy scales of primitive forms were of the type known as ganoid.

PALAEOPTERYGII, mainly extinct fishes, primitively with heterocercal tails and with the maxilla and preopercular bones as broad cheek-bones. They became very diverse in the Mesozoic era and some of the characters of the next group appeared singly in different species, but not in combination. The survivors are the sturgeons, degenerate in the loss of most of the bone from the internal skeleton, and the specialized bichirs (*Polypterus*,) with paddle-like pectoral fins.

NEOPTERYGII, including most of the present-day bony fishes. The tail-fin has an abbreviated axis ; the maxillary bone is freed from the cheek-plates and moves as the main bone of the upper jaw ; the preoperculum is boomerang-shaped, accommodating itself to the jaw-suspension apparatus beneath.

HOLOSTEI, the more archaic members of the Neopterygii, with several bones comprising the lower jaw, no supra-occipital bone in the skull and a series of valves counteracting back-pressure of blood as it leaves the heart. *Amia* and *Lepidosteus* survive.

TELEOSTEI, the modern branch of the Neopterygii, including nearly all familiar bony fishes. The lower jaw is simplified, a supraoccipital bone is present and the valves at the exit of the heart are reduced.

By using a natural system of classification of a group it is possible to show the affinities and derivations of its various sub-groups in the form of a diagram or evolutionary tree.

West side: second bay
Left-hand case

The Evolution of Man

One hundred years ago the view that man had evolved from lower animals was held by very few people; today it is widely accepted. In 1809 Lamarck stated his opinion that Man was descended from apes, but this revolutionary idea did not take root until after the publication of Darwin's *Origin of Species* in 1859. There the author showed that his theory of evolution would throw light on the origin of Man, and he developed this theme more fully in his *Descent of Man* published in 1871. At the present day the evolution of Man has come to be regarded as fact rather than hypothesis by all persons qualified to judge the relevant evidence.

As shown in other exhibits (p. 70), the resemblance in structure between different living organisms indicates the degree to which they are related, which is the basis of biological classification. Man is a warm-blooded, air-breathing animal with hairy skin, whose young are born alive and fed at the breast. He is therefore a mammal.

When Man's bodily structure is compared in detail with that of other mammals, it is clear that he belongs to the order of Primates, distinguished by large brains, and grasping hands and fingers with nails rather than claws. The Primates also include tree-shrews, lemurs, tarsiers, monkeys, and apes. Thomas Henry Huxley in one of his

Figure 12. The Evolution of Artistic Impulse. Photograph of a composition by the Chimpanzee 'Congo'.

By courtesy of Granada TV Network and the Zoological Society of London.

essays on *Man's Place in Nature* (1863), wrote : ' Whatever system of organs be studied . . . the structural differences which separate Man from the Gorilla and Chimpanzee are not so great as those which separate the Gorilla from the lower apes ' (i.e. monkeys).

The similarity between apes and Man also extends to behaviour including expression of emotion. Exceptional individuals among apes display the beginnings of artistic impulse as shown in pictorial compositions executed by chimpanzees.

The various Primates living today are the ends of divergent lines of evolution. By comparing them with one another and with fossil forms of earlier ages, palaeontological studies make it possible to picture their evolutionary relationship in the form of an ancestral tree (*shown also in the facing case of this bay*).

The Primates were derived from small insect-eating mammals similar to the Insectivores whose fossil remains have been found in Cretaceous rocks in North America. The living tree-shrews, closely related to Insectivores, represent a survival of the earliest type of Primate. Lemurs and the tarsier are also survivors of earlier types, all grouped as Prosimians.

Monkeys and the Hominoidea (apes and man) probably arose from advanced Prosimians. The earliest monkeys were not very different from tarsiers. The first apes were more monkey-like, and the earliest men more ape-like, than those of today.

The derivation of man from his ancestors was characterized by delay in development and preservation in the adult descendants of ancestral infantile characters. This mode of evolution, called paedomorphosis, is illustrated in another bay (p. 72).

In the embryological development of man there are many indications of his relationship with lower animals. At early stages the human embryo possesses visceral pouches and a tail and thereby resembles the embryo of fishes and other vertebrates (see p. 71).

Further evidence of relationship comes from serology. The serum precipitin test now even provides a means of establishing degrees of blood-relationship. Applied to the blood of Primates it has confirmed the conclusion that Man is related to the apes, and more distantly to monkeys (see p. 9). Man and apes share some of the same blood-groups and are subject to many of the same diseases.

On the evidence of comparative anatomy, of embryology and serology, zoologists now classify Man and the man-like or anthropoid apes as members of one super-family, the Hominoidea. This group is divided into two families, the Hominidae, which include the modern and extinct forms of Man, and the Pongidae, which includes the great apes and gibbons.

The conclusion that Man is related to the apes naturally implies that if their ancestries could be traced back far enough, they would be found to have diverged from the same stock ; it does not imply that Man has evolved from apes such as exist today. Fossil remains of Primates are rare, but those which have been discovered support this deduction.

Man's Ape-like Ancestors in the Miocene Period: *Proconsul*

The fossil which received the name *Proconsul* represents a primitive anthropoid ape, abundant in East Africa during early Miocene times, about 30 million years ago. The first specimen was found in 1926–7 at Koru on Lake Victoria. Further work around the north-east corner of the lake, particularly on Rusinga Island, by Dr. L. S. B. Leakey and others, led to the discovery of nearly 200 further specimens, representing three species of different sizes: *Proconsul africanus*, no larger than a modern gibbon; *P. nyanzae*, about the size of a small chimpanzee; and *P. major*, nearly as large as a gorilla. The specimens included one nearly complete skull, many jaw fragments and isolated teeth, and a few pieces of limbs including the greater part of a hand.

Proconsul was an early member of the family which includes the specialized 'great apes' of today, and its generalized, monkey-like characters indicate that it was close to or even part of the common stock from which the great apes and Man arose. Related forms living in Europe and Asia during Miocene and Pliocene times are known under the name *Dryopithecus*.

From the remains of *Proconsul* which have been found it is possible to form some idea of the probable appearance of this early type of ape. The middle-sized species portrayed in the drawing was more lightly built than a chimpanzee. It was an agile monkey-like creature, capable not only of climbing trees and swinging from bough to bough, but also of running along the ground on all fours. Whether, like all apes of today, it was tailless, is uncertain. Judging from the associated fossil plants and animals, *Proconsul* lived in tracts of forests affording a plentiful supply of fruits and other food. These forests extended along river courses, and were separated by bushy grassland. Volcanoes were active in the region, and it is thought that forest-fires resulting from volcanic eruption may have been responsible for occasionally driving these forest-dwellers into the open country where they became exposed to different conditions of natural selection.

The Transition from Ape to Man: *Australopithecus*

In 1924 Professor Raymond Dart obtained from a cave-deposit in a limestone quarry at Taung, Bechuanaland, the fossilized skull of a young ape-like creature which he named *Australopithecus*, meaning

Figure 13. Restoration of *Proconsul nyanzae*.

From a painting by Maurice Wilson.

'southern ape'. In some respects it appeared almost human, and was regarded by Professor Dart as a link between apes and Man. However, many authorities doubted whether this fossil was more closely related to Man than any other ape.

Between 1936 and 1951 the late Dr. Robert Broom found young and adult skulls, as well as other bones, of this and other allied forms in dolomite cave-deposits near Krugersdorp in the Transvaal. From the study of all the specimens now available, it is inferred that the Australopithecines, living in early Pleistocene times one million to half a million years ago, were early members of the family to which Man belongs.

The Australopithecines were small, about 4 feet in height, the later ones taller, with a brain scarcely exceeding in size that of a gorilla, but with teeth of essentially human type. Their backbone, hip girdle, and limbs were of human form and proportions; they walked upright on two legs, and, unlike apes, were able to live away from forests in dry open country, as is known from the fossil remains of other animals associated with them.

Life in the open country was precarious owing to the danger of attack from large carnivores, against which the Australopithecines had no natural defence, and their survival depended upon superior intelligence which was also required in meeting the need for catching animal food, to supplement the meagre vegetable diet available in grassland.

Even the earliest Australopithecines must have been able to make effective use of stones and sticks (or bones) with their hands which were no longer employed in locomotion. The discovery in 1959 by Dr. and Mrs. L. S. B. Leakey of pebble-tools in association with an Australopithecine skull on a lake-side occupation site at Olduvai in Tanganyika proved that the most advanced members of this group, living about 500,000 years ago, were tool-*makers*.

Earliest Men in Asia: *Pithecanthropus*

The earliest men in Asia were of the genus *Pithecanthropus*, known from Java and China. They were not only capable of tool-making, but of extending their range beyond the tropics by the use of fire and rock shelters; they were hunters and meat-eaters. These beginnings of culture were closely linked with expansion of the brain. The average brain-size of *Pithecanthropus* was twice that of *Australopithecus*.

While Eugène Dubois was searching for fossils in Pleistocene river

Figure 14. Restoration of *Australopithecus*.

From a painting by Maurice Wilson.

gravels in central Java in 1891, he found scattered fragments of a skeleton, including the top of the skull, isolated teeth, and a thigh bone, which he regarded as representing a form intermediate between apes and Man. He called it *Pithecanthropus erectus*, meaning the 'upright ape-man'. The name *Pithecanthropus* had already been proposed in 1866 by Ernst Haeckel for the hypothetical ' missing link '.

In 1937 Professor G. H. R. von Koenigswald obtained a second, more complete skull of the same species. From deposits of earlier age

Figure 15. Restoration of *Pithecanthropus pekinensis*.

From a painting by Maurice Wilson.

nearby he recovered parts of the skull of a related but more ape-like form and also the skull of an infant of the same species, *Pithecanthropus modjokertensis*. These early men with a cranial capacity ranging in the adult from 775 c.c. to 1000 c.c. lived in Java during Early and Middle Pleistocene times, between 500,000 and 200,000 years ago.

The thigh bone of *Pithecanthropus* was clearly that of an upright-walking man, and this confirms the indications already found in *Australopithecus* that Man's legs evolved in advance of this head which for a long time retained ape-like features such as the great overhanging brow-ridge.

In 1927, excavations in search of fossils in cave-deposits at Choukoutien, 30 miles south-west of Peking, led to the discovery of an isolated tooth recognized by Davidson Black as that of an early type of man, which he named *Sinanthropus* (man of China). Further excavations at Choukoutien, by Chinese and European scientists, resulted in the bringing to light of remains of more than 40 individuals of this extinct species of man, including 6 incomplete skulls, 12 jaws, 150 isolated teeth, and a few scraps of limb-bones. A comparative study of the Choukoutien skulls shows that they represented the same general type of man as the Javanese fossils, and for this reason, they are now included in the same genus but under a different specific name: *Pithecanthropus pekinensis*. The cranial capacity of this species ranged from 1000 to 1300 c.c.

The associated animals indicate that Peking Man lived early in Middle Pleistocene times, between 300,000 and 200,000 years ago.

The remains of Peking Man were found with ashes of fires, broken bones of the animals he hunted (mainly deer), crudely chipped tools of quartz and other hard stones, and the kernels of the cherry-like fruit *Celtis*. There is evidence that Peking Man was a cannibal with a predilection for human brains, for many skulls have been found broken open.

The implements and other remnants of the activities of early Man constitute what is known as a *culture*, the earliest of which are called Palaeolithic. Peking Man was in the Lower Palaeolithic stage of culture.

Right-hand case

The Emergence of Modern Man

Earliest Men in Europe

The oldest known trace of Man in Europe is the massive lower jaw found at Mauer near Heidelberg in 1907 at a depth of 82 feet in sand laid down by the river Neckar at the end of Lower Pleistocene times, about 300,000 years ago. The jaw is remarkable for its huge size and rather ape-like receding chin. The hind-part or ascending ramus is exceptionally broad, providing a large area for the attachment of the masseter muscle used in chewing. The top edge is not deeply notched as in the modern human jaw. The teeth are strongly implanted and have roots as large as their crowns, thus being well adapted to a coarse diet. No undoubted implements have been found with *Homo heidelbergensis*.

Three lower jaws similar in some respects to the Heidelberg jaw were found during 1953–5 in early Middle Pleistocene lake beds at Ternifine, Algeria, and were referred by Professor C. Arambourg to a new genus: *Atlanthropus*. They were associated with an early type of stone 'hand-axe'. The actual status of Ternifine Man, like that of Heidelberg Man, will remain doubtful until the greater part of a brain-case is found; but it appears likely that these men were members of a North African-European stock in process of emerging from the *Pithecanthropus* stage.

There are indications that by the end of Middle Pleistocene times, in Europe at least, men had passed beyond the *Pithecanthropus* stage and included variable types constituting the stock from which emerged both *Homo sapiens* and the less successful *H. neanderthalensis*. The skull found in 1933 at Steinheim near Stuttgart, in river sand of second interglacial age, is typical of this early 'pre-neanderthal' or 'pre-sapiens' group. The Steinheim skull has fairly heavy brow-ridges and a strongly built upper jaw with large teeth; but the vault is moderately high and the back is well filled out as in *Homo sapiens*.

In 1935 and 1936 Mr. A. T. Marston discovered two fragments of a human skull in the gravels of the 100-foot terrace of the Thames at Swanscombe, Kent. They were the *occipital* bone (forming the back and part of the base of the skull), and the left *parietal* bone (which forms part of its roof). The two pieces, which fitted together perfectly, were found 8 yards apart but in the same seam of gravel, 24 feet below the surface. In 1955 excavations by Messrs. B. O. and J. Wymer and A. Gibson revealed the right parietal.

The skull was associated with Acheulian flint 'hand-axes' and with the remains of animals, some now extinct, such as the elephant *Elephas antiquus*, which inhabited the woodlands bordering the Thames during the second interglacial period, 200,000 to 100,000 years ago. Fluorine, nitrogen, and uranium tests confirmed that the skull and animal remains were contemporaneous.

Apart from the unusual thickness, the parts of the Swanscombe skull preserved do not differ from the corresponding bones in some modern crania. The brain-case had a capacity of about 1325 c.c., close to the average in modern Man. The nasal air sinuses extended very far back, suggesting that the face and jaws were of heavy build. As the frontal bone of Swanscombe Man has not yet been found, it is uncertain whether the brows were ridgeless as in the Fontéchevade Skull and in modern Man, or had overhanging brow-ridges as in the contemporaneous Steinheim Skull. Even without knowing how the frontal bone of

Figure 16. Restoration of Swanscombe Man.
From a painting by Maurice Wilson.

Swanscombe Man was shaped, it is nevertheless certain that it did not conform to the specialized type of Neanderthal Man.

Neanderthal Man

The first fossil human skull to be discovered came to light in a cave during quarrying on the North Face of the Rock of Gibraltar in 1848. It was not recognized as that of an extinct type of man until another example was found in a cave in the Neanderthal valley near Düsseldorf in Germany in 1856, and this was confirmed when several skulls of the same type were found at Spy in Belgium. The discovery of the

Neanderthal skull was recognized at the time to be a striking confirmation of Darwin's application of the theory of evolution to Man.

Remains of more than 60 individuals of *Homo neanderthalensis* have since been found at widespread localities in Europe, Western Asia, and North Africa. Neanderthal skulls have prominent brow-ridges which give them an ape-like appearance, and were low-crowned and flattened towards the back. Neanderthal Men were stocky, about 5 feet high. Their legs were rather short with the thigh-bone bowed forward.

The typical Neanderthal Men were cave-dwellers. They lived in Western Europe under the severe climatic conditions accompanying the first stages of the last glaciation, 70,000 to 40,000 years ago. In physical evolution they had diverged from their more progressive cousins of the Mount Carmel Caves (about to be mentioned), and after a period of predominance they died out.

The culture of Neanderthal Man is named Mousterian, after Le Moustier near Peyzac, in the Dordogne. Mousterian stone implements were more specialized than those of earlier cultures, so that this culture is classified as Middle Palaeolithic. The principal tools were flakes with trimmed edges, points for piercing and cutting, and D-shaped side-scrapers believed to have been used in preparing skins for use as wraps. The Neanderthal people increased their adaptation to the glacial climate by the regular use of fire, and by using rock-shelters as dwelling-places. They hunted game with the aid of wooden spears and balls of stone or hardened clay used as missiles. They practised cannibalism, and occasionally at least buried their dead with some ceremony.

Neanderthal-like Peoples in Asia and Africa

The typical Neanderthal Men were confined to Western Europe, but there was a closely related stock in Western Asia, represented by Mount Carmel Man. In the Far East and in Africa there were similar types such as the Solo Man of Java, Saldanha Man of South Africa, and Broken Hill (Rhodesian) Man of Central Africa. All these extinct forms of Man resembling but not identical with Neanderthal Man are called Neanderthaloids.

Remains of more than a dozen Neanderthaloids have been found in the caves of Palestine, notably in caves named Tabun and Skhul in the Wady el-Mughara (' Valley of Caves ') on the western side of Mount Carmel.

The Palestinian Neanderthaloids were remarkable for their variability which was greater than in any community of modern times. They ranged, from individuals with almost typical Neanderthal skulls such as were found at Tabun, to others found in the Skhul Cave which were barely distinguishable from modern Man (*Homo sapiens*).

The men were tall, up to 5 feet 10 inches in height, with long straight legs and walked upright as easily as Man of today. Their skulls, although having brow-ridges, were well domed. Neanderthal features were more marked in the women.

Figure 17. Restoration of Neanderthal Man at Gibraltar.
From a painting by Maurice Wilson.

Mount Carmel people lived at a time when the climate of Palestine was changing through the onset of the last glaciation. They appear to have been progressive Neanderthaloids in process of modification or assimilation into *Homo sapiens*. They were hunters and gatherers of wild foods ; they used fire, and they buried their dead. Their implements were a variety of Mousterian.

In 1921 a fossil human skull of brutish appearance was found during the mining of lead and zinc ores at Broken Hill in Northern Rhodesia. Other human bones were also found. These remains were deep in the mineralized deposits of a cave containing quantities of broken animal bones, and Palaeolithic stone and bone implements, dating from late Pleistocene period, between 50,000 and 20,000 years ago.

The skull was described by Smith Woodward as that of an extinct species, *Homo rhodesiensis*. It is probably the skull of a male about 40 years old, and is remarkable for the heavy brow-ridges and the large size of the face and palate. The brain-case is not much smaller than the average in Man of today. From the position of the opening for the spinal cord it is evident that the head was held upright. The neck muscles were powerful.

The owner of the skull when alive suffered from severe dental decay which is very rarely found in prehistoric man; and judging by his shin-bone he also suffered from arthritis.

A skull of the same type was discovered near Saldanha Bay, South Africa, in 1953, where it was associated with a very late Acheulian industry.

Rhodesian and Saldanha Man were the South and Central African equivalents of Neanderthal Man, and were replaced or absorbed by more adaptable variants of *Homo sapiens*, including Proto-Bushmen represented by the Singa Skull from the Sudan (exhibited in the Fossil Mammal Gallery).

Upper Palaeolithic Man in Europe

In Upper Pleistocene times, during a temporary amelioration of the climate, beginning about 40,000 years ago, the Neanderthal Men in North-west Europe were replaced by people of entirely modern type, *Homo sapiens*. These newcomers probably originated in South-west Asia, where they had developed a distinctive culture known as Aurignacian. There are indications in Syria and Palestine that Aurignacian culture was foreshadowed in that area while the Mount Carmel people were making Mousterian implements.

Remains of earliest undoubted representatives of *Homo sapiens* were first discovered in the rock-shelter of Cro-Magnon at Les Eyzies in south-western France, in 1868. The typical Cro-Magnons were finely built, tall, the men sometimes over 6 feet in height, muscular, with high domed skulls and refined facial features. Several variants occurred, notably the Combe-Capelle or Predomost type, with larger brow-ridges; the Grimaldi type, of negroid appearance; and the Chancelade type, with some resemblance to the Eskimo.

In France three main cultures have been recognized in deposits left by *Homo sapiens* in caves and in the open during the various stages of the last glacial period. These cultures are classed as Upper Palaeolithic and have been named as follows, the oldest being the lowest:

3. Magdalenian (after La Madeleine, Dordogne).
2. Solutrean (after Solutré, Mâcon).
1. Aurignacian (after Aurignac, Haute Garonne).

The people with Aurignacian culture were mainly Cro-Magnon and Combe-Capelle types, though a few were of the Grimaldi type. Solutrean and Magdalenian communities included individuals of the Cro-Magnon and Chancelade types.

Upper Palaeolithic people were more elaborately equipped than their predecessors; they worked bone, their flint tools were delicately worked blades, they used spears and harpoons with detachable heads, invented a spear-throwing device, made sewn clothes of skin or fur, and in open country they constructed tents and huts. Above all, they are notable for their artistic achievements, such as the cave-paintings of Lascaux, probably connected with some hunting ritual. They adorned themselves, for example, with necklaces made of pierced shells, animals' teeth, and occasionally amber. Like all Palaeolithic peoples, they did not cultivate plants or domesticate animals, but obtained their food by hunting wild animals, particularly bison, horse, reindeer, and mammoth, and by gathering edible plants.

Although the Caucasoid (e.g. European), Australoid, Mongoloid, Negroid and Bushmanoid types of men were foreshadowed in Upper Palaeolithic populations, the modern races of Man were not fully differentiated until early Post-glacial times, in the Mesolithic and Neolithic stages of culture.

West side: third bay
Left-hand case

Man as Hunter: Palaeolithic Stages of Culture

Throughout the Pleistocene period Man lived by gathering wild foods and hunting other animals. His survival in a wide range of environments depended on his ability to reason, to communicate and store experience, and to make tools and weapons suited to his needs. A basic requirement was the possession of tools for cutting through fur and skin, and for shaping sticks. At the dawn of culture, sharp pieces of stone casually picked up served such purposes; but as time went on man learnt to break stones deliberately and to shape them with foresight of his precise needs.

The immensely long periods during which man depended on deliberately shaped stone cutting tools is called the Stone Age. That part which coincided with the Pleistocene period is called the Old Stone Age, or Palaeolithic Period (500,000 to 10,000 years ago).

Traces of man in Europe are rare before the second or maximum glaciation. During the succeeding warm interglacial period, groups of early Palaeolithic Man spread widely. Some became inhabitants of southern Britain, which was then joined to the Continent.

Clacton: An Early Palaeolithic Site

During the early part of the Pleistocene period, the Thames entered the North Sea at a point north-east of the present mouth of the River Blackwater. An abandoned channel of the early Thames now outcrops on the foreshore of Clacton-on-Sea, Essex. It contains gravels, sands and loams deposited by the river during the interglacial period when man was an established hunter in the river valleys of southern Britain.

The Clacton 'Elephant Bed', as this ancient river deposit has been called, was discovered by John Brown of Stanway about 1838. It has since been investigated by Mr. S. Hazzledine Warren, whose collection from the site is exhibited. Evidence of a very primitive flint industry occurs in this ancient river bed. It consists of flake-tools and choppers, and represents a cultural tradition called Clactonian, which appears to have been distinct from the tradition of the contemporaneous makers of Acheulian flint hand-axes.

It is probable that Palaeolithic man made extensive use of wood for tools and utensils which have usually vanished without trace through decay. Many of the flint scrapers of the Clacton industry would have served for shaping spears and other wood-work. The broken spear made of yew-wood (*Taxus baccata*) found in the Clacton deposits in 1911 is the oldest known wooden artifact. It was embedded in a peaty loam and it owes its preservation to the fact that the deposits are at sea-level and have been water-logged since the time of their formation, roughly about 150,000 years ago.

From the seeds of plants, shells of molluscs, and bones and teeth of mammals found in these deposits, it is possible to build a picture of the flora, fauna, and climate of southern Britain at the time when the Clactonian hunters lived on the banks of the Thames.

Study of the seeds and pollen-grains preserved in the silty layers of the Clacton deposits has established that the climate was temperate and that mixed-oak forest covered much of the surrounding country. Pollen-analysis has established further that the changes that occurred in the vegetation were those characteristic of the warm second interglacial period. The Clacton deposits accumulated during the same period as the lake-beds at Hoxne in Suffolk, where John Frere made the discovery of stone implements in 1797. The Hoxne implements are hand-axes similar to those found at Swanscombe, and named Acheulian, after St. Acheul, near Amiens in northern France. The Acheulian culture was perhaps adapted to a less wooded environment than the Clactonian

The Clacton deposits contain bones and teeth of the wild animals that browsed and grazed in the woodlands and water-meadows bordering the river. They included extinct fallow deer, wild oxen, rhinoceros related to the living Sumatran form, and elephants which belonged to the extinct species *Elephas antiquus*, characteristic of a warm climate. Small herds of horse and bison roamed over the scrubby grassland beyond the woods. All these were preyed upon by the much rarer carnivores such as lion and bear, and by man who probably ambushed and killed them when they came down to the river to drink. Other mammals recorded in the Clacton deposits include hyaena and beaver.

Facing case

The Spread of Man

The earliest members of the Hominidae included the first men or tool-makers originated in the tropics, probably in the dry grasslands and savannahs bordering the forests of Central Africa, whence they spread widely, through the use of equipment which enabled them to survive in environments to which they were unfitted by nature. A map shows those areas of the Old World which had been entered by man during the Early Palaeolithic stage 500,000 to 50,000 years ago, and how much further man had spread by the Mesolithic stage 10,000 years ago.

In early Palaeolithic times man had spread all over Africa, northwards in Europe up to the latitude of Doncaster in Britain, over the southern lowlands of Asia up to the latitude of Peking, and through South-eastern Asia to Java which was then connected with the mainland.

By the end of Upper Palaeolithic times, Man had become sufficiently adapted to cold conditions to break the bounds of the Old World by crossing the Bering Straits land-bridge into Alaska. By the time of the Mesolithic settlement at Star Carr in Yorkshire about 7500 B.C. Man had spread through the grasslands of North America, and down the western margins of South America to the southernmost tip of that continent. The spread of Mesolithic man can be described as circum-Pacific, for the radio-carbon dating of the earliest remains of aborigines in Australia indicates that they had also reached that continent by that date.

Right-hand case

Man as Hunter: Mesolithic Stage of Culture

After the change in climate which ended the Ice Age, about 8000 B.C., a variable period still elapsed before men took to farming. During this interval, much longer in Europe than in the Middle East, men still lived mainly by hunting and fishing, but they were in the adaptable stage of culture known as Mesolithic.

At the beginning of Post-glacial times, Britain still formed part of the Continent, and until the rising waters flooded over the southern part of the North Sea bed about 6000 B.C., it was still possible to walk from East Yorkshire to Esthonia and beyond, crossing rivers and skirting bogs and lakes. Examination of samples of peat ('moorlog') dredged from the Dogger Bank show that at the time of the submergence, this was part of a great fenland with woods of birch, willow, and pine.

At this time the heavily wooded North European plain was occupied by Mesolithic hunter-fishers who succeeded the Upper Palaeolithic reindeer hunters. Many of their camping sites have been found on the Continent, notably in Denmark, and from their habit of camping during the summer months on the margins of lakes and bogs, these early forest dwellers have come to be known as Maglemosians (Danish: *magle* = great, *mose* = moss or bog). Their mode of life is fairly well known, for in peat and waterlogged deposits their material equipment has been preserved.

Star Carr: a Mesolithic Site

Scattered finds have for long indicated that the low-lying parts of eastern England were occupied by scattered groups of Maglemosians, but no peat-bog site comparable with those of the Continent had been found until the researches of Mr. John Moore led to the discovery in 1948 of the relics of a red deer hunters' settlement below peat at Star Carr near Seamer in Yorkshire.

During the summers of 1949–51, large-scale excavations were undertaken at Star Carr by the Prehistoric Society, under the direction of Professor Grahame Clark. The site proved remarkably rich, particularly in organic material. From pollen analysis of the peat and examination of associated brushwood, the contemporary flora can be deduced; from a study of the numerous animal bones the fauna on which these hunters depended—red deer, elk, and wild oxen—is known; while from the abundant archaeological material it has been possible to reconstruct the way of life in Britain at a time which radio-carbon dating of material from the site indicates was about 7500 B.C.

The excavation revealed a brushwood platform of a most primitive character, weighed down by stones, and showing traces of birch bark flooring and of charring of the wood by domestic fires. There were no indications of the nature of the super-structures, but they may well have been tents.

The dwelling site was close to the shore of the dwindling glacial Lake Pickering, and the ground would have been marshy and perhaps liable to flooding. The construction of dwellings on brushwood platforms over marshy ground was a step leading in later prehistoric times to the building of artificial islands such as the crannogs in Ireland, the Glastonbury 'lake-village', and other forms of lake-side villages.

The chief weapons of the Maglemosian hunters were spears or arrows with barbed antler points. More than 50 of these points were found in the excavation of an 8-yard trench at Star Carr, many of them complete. Previous to this discovery, only 6 were known from the British Isles, including one dredged from the floor of the North Sea, embedded in peat.

At one time these barbed points were called harpoons, but this term is not generally applicable because only a few were detachable weapon-heads joined to the haft by a line. It is probable that those points with a long basal part were hafted in pairs as leister prongs for spearing fish, in the way that eels are speared in the West Country at the present time. Some of the points barbed almost to the base may have been used as the heads of arrows or darts in hunting deer. The bow had certainly been invented by this time, for at least one bow-stave has been found in a Maglemosian level in a Danish bog.

Antler was the principal raw material of the Mesolithic people of Star Carr. They worked it with flint tools, chiefly the chisel-pointed implements known as *burins*, which were abundant on the site. The techniques employed had been inherited as a tradition from the Upper Palaeolithic reindeer-hunting tribes.

The character of a Mesolithic culture was largely determined by the natural environment, to which the Maglemosians of eastern England, for example, skilfully adapted themselves. The dominant animal there was red deer, and the local woods were mainly of birch. Consequently the material equipment of the early settlers at Star Carr was mainly fashioned from red-deer antler, birch wood and birch bark.

In Britain, Man was in the Mesolithic stage until 2500 B.C., while in Australia the aborigines have remained in this stage until the present day.

West side: fourth bay
Left-hand case

Man as Farmer: the 'Neolithic Revolution'

For half a million years or more, in the Older Stone Ages (Palaeolithic and Mesolithic), all mankind lived by *food-gathering*, that is to say, by collecting wild plant foods and hunting animals. In the New Stone Age or Neolithic stage, starting in the Near East probably between 8000 and 7000 B.C. (but in Britain much later, about 2500 B.C.), Man began *food-producing*, by agriculture and animal husbandry. Farming revolutionized Man's way of life, enabling him to live in settled communities and to begin pursuing the arts of civilization.

The principal plants cultivated by Neolithic Man in Western Asia and Europe were cereal grasses (wheat and barley), pulses (peas and beans), flax (for supplying oil and fibres), and fruit (for example, apples). The advantages of cultivating grasses were probably first realized by groups of food-gatherers living a fairly static existence on the grassy uplands bordering the arid plains of the Near East, for example near Mount Carmel. Within this region grew several annual grasses, notably the wild wheats einkorn and emmer, and wild barley, tolerant of low rainfall and yielding large enough grains in sufficient quantities to encourage harvesting (see p. 54).

Just as the civilization of Western Asia and Europe arose on the basis of the cultivation of wheat, those of the Far East and Central America were based on rice and maize respectively. All these cereals contain a concentration of energy-producing carbohydrate, and all are native to areas where protein-rich pulses are also found wild.

The wild ancestors of domestic dogs, goats, sheep, cattle, and pigs, existed in the areas of the Near East where wheat and barley were brought under cultivation, and it is probable that they were domesticated there: dogs in the Mesolithic stage, goats and sheep at the beginning of the Neolithic, cattle and pigs slightly later. The domestication of these animals is unlikely to have begun as a purposeful procedure, but rather as a result of natural association with man. Wild dogs must have entered human settlements as scavengers, and the puppies would have been secured as pets by the children. Tamed dogs would have helped the Mesolithic hunters in driving sheep and goats; and with the beginning of agriculture, wild cattle and pigs would have become associated with human settlements when they raided the crops.

As food supplies increased and became assured through Neolithic economy, not only was a more settled mode of life made possible, but specialized craftsmen and traders could be supported. A wide variety of new materials were used. Some early Neolithic people were without pottery, but used leather vessels, baskets and carved stone bowls. The forms and decoration of Neolithic pottery were evidently influenced by leather, basket, or stone prototypes. Other new crafts and materials included the spinning and weaving of plant and animal fibres to use in clothing and in furnishing dwellings.

The religious ideas of Neolithic Man were related partly to fertility, as is shown by statuettes of Earth Mother goddesses and phallic emblems, and partly to ancestor worship which led to the construction of large stone tombs. The Neolithic stage is distinguished by the first permanent villages; and the exceptional spring-side settlement now being excavated at Jericho was a veritable town by 6800 B.C. according to the latest radio-carbon age estimation. The activities of Neolithic Man included the continuation of hunting and fishing using new devices, the mining of flint, the manufacture and trading of polished stone axe-heads, and their use in the clearance of forests for agriculture.

During an experiment in Denmark, three men cleared 600 square yards of forest in four hours using a Neolithic axe. More than 100 trees were felled with one axe-head, which had not been sharpened for more than 4000 years. A pollen-diagram based on samples taken from borings in Danish peat-bogs shows the effects of clearing forests for agriculture about 2500 B.C. The great increase in birch pollen following the disappearance of the big trees suggests that the clearings were burned in readiness for sowing. The return of big trees indicates that after a time the ground became exhausted, and the Neolithic farmers had to move on to new ground.

Right-hand case

The Races of Man

It has often been debated whether the living races of man can be regarded as distinct species ; but since no detectable degree of sterility results from race-crossing, it is now agreed that they are all one species, *Homo sapiens*, comprising at present about 2,500,000,000 individuals. The simplest, but arbitrary, classification of mankind is into three major divisions : the ' white ', ' black ', and ' yellow ' races, but sub-division into thirty or more races has also been attempted.

The visible and measurable characters generally used in race classification (skin colour, hair form, shape of head, stature) are no longer regarded as wholly reliable guides to relationship. They do not allow clear-cut divisions since they are controlled by many genes with cumulative effects. It is probable that the same characters may have appeared independently during the evolution of different races.

A race may be defined as an inter-breeding population differing from other populations within the species in the incidence of certain genes. The discovery that some of the biochemical properties of blood are dependent on single genes has made it possible to begin classifying human populations on a genetic basis. By taking into account the differences in the percentages of blood-groups in different typical populations (see Table 1) W. C. Boyd has recognized six genetically defined races : Early European, European, Australoid, Negroid, Mongoloid, and Amerindian.

Races are not static entities. The genetic constitution of a population is continually changing through mutation, cross-breeding, natural, and sexual selection. In man there is in addition the mechanism of psycho-social evolution, which has accelerated racial change. Existing races of mankind are of comparatively recent origin; they were barely differentiated as the Ice Age was ending about ten thousand years ago.

Table of Frequencies of Blood Groups in Typical Populations of Man

	Blood group O	A	B	AB	Other distinguishing factors
	(percentages)				
Early Mediterranean:					
Basques	57	42	1	0	Highest Rhesus-negative
Caucasoid:					
Western European	46	42	9	3	
Negroid:					
Senegalese	51	23	24	2	High Rhesus $_0$
American Negroes	48	26	23	3	High Rhesus $_0$
Bushmen	54	30	9	5	High Rhesus $_0$
Mongoloid:					
Chinese	31	25	34	10	
Amerindian:					
Navaho	78	22	0	0	High Rhesus $_z$
Australoid:					
Australian Aboriginal	54	41	4	1	High N

(Based on researches by A. E. Mourant, E. B. Ford, W. C. Boyd.)

The exhibit shows the main living variants of the human species, grouped as far as possible on the basis of Boyd's classification, and their prehistoric precursors.

PANEL I: AUSTRALOIDS AND OCEANIANS

Apart from their darker skins and distinctive blood-group pattern (high percentage of blood-group N), the Australian aborigines fall within the range of variation of the more primitive Whites, to whom they seem closely related. They migrated from Asia at the end of the Ice Age. At that time the sea level was low, and the sea channels between Asia and Australia were fewer and narrow enough to be crossed on rafts. The first wave of migrants were of Melanesian type, represented by the Tasmanians (who became extinct in 1876). They were followed by the more typical Australian aborigines who presumably had boats and brought dogs with them. Two main strains can be distinguished, namely the Murrayians, stocky and with abundant body hair, and the

Carpentarians, who are taller and leaner. The Veddoids of Asia possibly represent the ancestral stock of the Carpentarians, and the Andamanese may be a dwarfed residue of the stock from which the Melanesian wave came. The hairy Ainu of Japan may be related to the ancestral Murrayian stock.

Where the tall, light brown Polynesians came from is still unknown, but they form a homogeneous group, and reached the islands less than 1500 years ago in a Neolithic state of culture. Although cultural evidence points to a western origin, in their blood-groups they closely resemble the aborigines of North-west America. This has been used to support the hypothesis of Thor Heyerdahl that Polynesia was populated by migrations across the Pacific Ocean from the east, the first from Peru, the second from North-west America.

The dark-skinned peoples scattered through the islands and archipelagoes of South-east Asia and Oceania had a different origin. They inclued pygmies (Negritos) and others, such as the Melanesians, who are of medium height. They resemble the Negroes of Africa, particularly in hair and in skin-colour but their blood-groups show that they are not of African origin. These Negroid features are evidently adaptations to bright light and intense heat.

PANEL 2 : MONGOLOIDS AND AMERINDIANS

The Mongoloids have yellowish or brownish skins, straight black hair, little or no hair on face and body, rounded heads, broad flat faces with high, flaring cheek-bones, and a characteristic fold of skin over the upper eye-lid. This physical type appears to have evolved in North-east Asia, probably as a result of the population there being isolated during the last part of the Ice Age, and subjected to intense selection of the characteristics which aid survival in extreme cold. Thus, the eyes of the Mongoloid are padded with fat, the cheeks are built out with fat-pads to form a flat surface, the nostrils are small, the frontal ridges and associated sinuses (vulnerable to cold) are much reduced, while the yellow-brown skin colour and the reduced size of the eye-opening are natural protections against snow-glare.

As the original Mongoloids, essentially nomadic people, increased numerically, they spilled over into adjoining lands whenever conditions allowed. Mongoloid populations are distinguished by having a very high incidence of the blood-group B gene.

The Amerindians, aborigines of the Americas, are closely related to the Mongoloids. Their ancestors migrated from Asia across the Bering

Straits during the closing stages of the Ice Age. The earliest waves of migrants were probably close in type to primitive Whites or Australoids, later waves were more Mongoloid in appearance. The absence of blood-group B from the majority of Amerindian populations (excluding the Eskimoes) indicates that the migrations took place mainly before the great increase in frequency of blood-group B occurred in Asia.

In Central America the aborigines have survived in tropical heat and sunshine for at least 10,000 years without developing black skins. It has been suggested that as their ancestors passed through the 'cold filter' of Siberia the genes required for the development of black skin were eliminated.

The Eskimoes are the most recent arrivals of all the aborigines of the Americas. They stand out strikingly from other Amerindians in speaking languages of a single group, all having a similar fishing and sealing culture, and being specialized in adaptation to an Arctic environment. They are more like the Mongoloids of Siberia than are any other inhabitants of North America, but in showing a very high frequency of the blood-group M gene, relatively to N, the Eskimoes conform with Amerindians. The fairly high frequencies of blood-group B among Eskimoes indicates, however, their comparatively recent derivation from Asiatic stock.

PANEL 3 : NEGROIDS

North of latitude 15 °N., Africa is inhabited mainly by the African Whites, or Hamites, with the light-brown skin characteristics of the Mediterranean peoples; but to the south of that latitude the native inhabitants are of the 'Negroid Race' including Negroes, Pygmies, Bushmen, and Hottentots, characterized by spiral hair, and a high incidence of the Rh_O blood-group gene. The true Negro is mainly confined to West and Central Africa, for in the east and south the Negroes have been modified by mixture with Hamites to form, for example, the Nilotes and the Bantu. The Pygmies, with mahogany-coloured skin, occur in the Congo rain forests.

The Bushmen, of whom there are still nearly 50,000 living in small bands scattered through the Kalahari Desert and parts of South-west Africa and Angola, are hunters and food-gatherers. They are short and have yellowish-brown skins toning with the desert background. The women show exceptional development of the buttocks (steatopygia).

The Hottentots are physically similar to Bushmen, but taller and with narrower heads. They originated in North-east Africa, probably through hybridization between Early Bushmen and Hamites. They are pastoralists with herds of cattle and fat-tailed sheep. Southern Africa was at one time wholly occupied by Bushmen and Hottentots, but from the 17th century onwards they were overrun by the more successful, black-skinned Negroes, spreading in large numbers from Central Africa, and they have only survived in the less hospitable areas. The Cape coloured people have a large Hottentot component.

The main characteristics of the Negro are the deeply pigmented skin, hair and eyes, woolly hair, relative tallness, long head, flat broad nose, thick everted lips, and prominent jaws (prognathism). There is a paucity of hair on the body, and an abundance of sweat glands, which by their secretion and the consequent evaporation of sweat keep the body cool.

Where sunshine is very intense, an injurious amount of light (particularly ultra-violet) reaches the interior of the eye, unless the iris is deeply pigmented. In Negroes, eye-colour ranges from dark brown to black. In addition to a layer of dense pigment on the surface of the iris, there is a layer of melanin granules at the base of the retina which acts as an anti-glare mechanism and thus increases the resolving power of the eye in brilliant light. The ability to obtain a very sharp visual image of distant objects was important for the survival of men hunting animals in glaring desert or grassland.

PANEL 4: CAUCASOIDS

The inhabitants of Europe and adjoining parts of Asia and Africa share a number of physical characteristics which allow them to be grouped as Whites or Caucasoids. In skin-colour the Caucasoids range from white to brown, but their common characteristics include hair that is fine, wavy and strongly developed on the face in males, noses that are rather narrow and high-bridged, thin lips and prominent chins. On the basis of appearance and measurements, four main types can be distinguished. *Nordic:* tall, blond, with light eyes, long and rather large head, long face, deep chin. *Mediterranean:* shorter in stature, long head but usually of more delicate build, brunette complexion, dark eyes and hair sometimes curly. *Alpine:* medium height, thick-set, sallow complexion, round-headed, broad forehead, nose of medium breadth often with blobby tip. *Dinaric:* taller, broad but high head with narrow forehead and prominent convex nose.

On the basis of blood-group evidence, a European race can be defined genetically, but this excludes some populations, particularly in North Africa and West Asia, which although conforming to the Mediterranean type are evidently of mixed origin.

The Basques and perhaps the Berbers have a distinctive blood-group pattern, representing an *Early Mediterranean* race. The Basques, of whom there are about three quarters of a million living in Northern Spain and the adjoining part of South-western France, are distinguished from all other inhabitants of Europe by (1) having the highest-known incidence of the Rhesus-negative blood-group gene in the population, (2) having the lowest incidence of the blood-group B gene, and (3) speaking a pre-Indoeuropean language. They are evidently survivors of an Early Mediterranean people who spread from the East before 2000 B.C. and included the Ligurians (de Beer).

People of Mediterranean type predominate in the countries around the sea from which they derive their name, and include the Hamites of North Africa, the Semites, Arabs, Persians and lighter-skinned inhabitants of India and Pakistan (largely descended from the Aryans who brought Indoeuropean speech into India about 3500 years ago). They vary regionally. In parts of Asia they have convex noses and strong beards; in the west their noses are straighter and beards less strong. In Europe the Mediterraneans have mostly blended with Alpine, Dinaric, and Nordic types. The Jews, who originated in the Levant, and the Gipsies, who originated in North-west India (as confirmed by blood-group evidence), have preserved the characteristics developed in their homelands through the operation of social forces, discouraging exogamy.

The Neolithic immigrants in Western Europe were of Mediterranean stock. They were mainly long-headed (dolichocephalic), but in some of the aboriginal Mesolithic communities round-headed individuals were already quite numerous (e.g. at Ofnet in South Germany 8 out of 21 complete skulls in a Mesolithic mass-burial were of rounded form). In the Bronze Age there were extensive movements of predominantly round-headed people from Central Europe, Spain, and Asia Minor. The factors responsible for brachycephaly are unknown, but since the skull is slightly broader in infancy than in maturity, it may be an example of paedomorphism or persistence of infantile characteristics in the adult stages. (See p. 72.)

Blondness is the most distinctive physical trait shown by Europeans. Among the predominantly brunette Mediterraneans, blond individuals are in a small minority but, further north blondness increases, reaching a maximum in the Baltic region.

Fair-haired members of the Celtic language-speaking groups, which form an important Nordic component of the population of the British Isles, contrast with the swarthy, black-haired stock of ultimately Mediterranean origin, recognizable particularly in the populations of Eire, Wales, Cornwall, and Brittany. According to the evidence of Classical writers, light-haired individuals were frequent amongst some of the Celtic tribes in Early Iron Age times. Red hair certainly appears to be commonest amongst the modern populations of ' Celtic ' origin.

THE SIX MILESTONES OF EVOLUTION

Life; Plants and Animals; Sex; Multicellular Organisms, Embryonic Development, and Death; the Colonization of Land; and Psycho-social Evolution

During the evolutionary history of the earth, beginning with molecules and leading to man by the processes of variation and selection, there have been six major steps initiating new stages of organization which have had lasting effects. First was the production from inorganic molecules of self-regulating, self-copying, and self-reproducing systems ultimately resulting in organisms in the form of living cells, such as bacteria. This was the evolution of Life. Next came a distinction between those organisms which by means of chlorophyll utilizing sunshine continued to live on inorganic raw materials, the plants, and those which took to feeding exclusively on other organisms, the animals. A step of fundamental importance was taken when organisms evolved the habit of pooling and exchanging their hereditary units the genes, and recombining them in different ways at the formation of new individuals. This reshuffling of genetic material is the basis of sex, and its most important feature was that an immensely greater amount of variation became available on which natural selection could act. The fourth great step in evolution was the transition from the single-celled to the many-celled condition, a change which brought with it two consequences: embryonic development of the new generation to convert it from a single cell into a many-celled structure similar to that of its parents, and, accompanying this step, the inevitable natural death of the parents. Death is the price paid in each generation for the increased complexity of body and possibilities of life resulting from the luxury of being composed of many cells. Fifthly the colonization of land opened a vast new range of possibilities for the evolution of the higher plants and animals. Finally there was the inception of psycho-social evolution exemplified only in man, whose higher mental faculties, with ability to exchange experience by articulate speech, store it in memory and impart it by tradition, enable him to choose lines of conduct and control his own further evolution.

West side: fifth bay
Left-hand case

The Origin of Life

At the early stages of the earth's history Life was impossible because of the physical and chemical conditions that then prevailed, whether the earth cooled from a molten state or concentrated out of cold cosmic dust. Life on earth must therefore have started at some time in its history ; but it is probable that the physical and chemical conditions which prevailed when life originated have not persisted, and that new living organisms have not arisen from non-living matter since that time.

It may be supposed that in the primeval oceans, salts became concentrated and that with the energy supplied by the sun they were built up into increasingly complex molecules.

Since Kolbe synthesized acetic acid, and Berthelot synthesized hydrocarbons such as acetylene and obtained many derivatives from them, it has been clear than many so-called organic chemical substances found in living organisms can be prepared in the laboratory from their constituent elements, and that organisms must originally have been evolved from compounds found in nature.

Amino-acids, of which glycine is an example, are the basic building materials of which proteins are composed, and therefore essential ingredients of living matter. Amino-acids can be synthesized by the action of electrical discharges and ultra-violet light on mixtures of simple substances such as hydrogen, water, carbon dioxide, methane, and ammonia. (S. L. Miller.)

Typical proteins consist of complex chains of amino-acids forming polypeptides.

Adenosine triphosphate is a substance which plays a very important part in the reactions of living matter because it can transfer free energy from some molecules to others, thereby furnishing the energy required for muscular contraction, electrical discharge in certain fishes, light-production in fireflies, and the synthesis of complex biological materials. In so doing, ' A.T.P. ' parts with one or two of its phosphate groups ;

but these can subsequently be replaced by means of energy provided from the oxidation of other substances. The original molecule of A.T.P. can thus be regenerated.

Deoxyribose nucleic acid, of which a molecular model is shown, consists of two spiral chains of phosphate and sugar groups, interconnected at intervals by nucleotide bases in pairs, consisting of either guanine and cytosine, or of adenine and thymine.

'D.N.A.' is believed to be the chemical basis of the genes, large numbers of which, together with proteins, constitute the chromosomes. When the molecule of D.N.A. uncoils and splits, each portion is incomplete and takes up what it lacks from the available materials in the cell to produce two copies from the original single structure. This is the basis of self-copying and reproduction. Sometimes the copying is not exact and the nucleotide bases in one of the products are differently arranged. This is believed to be the basis of mutation. (J. D. Watson, F. H. C. Crick, and M. H. F. Wilkins.)

The formation of new compounds may be regarded as 'variation', and selection must already have been at work before life began, for unless the molecules were to a certain extent stable, those particular chemical compounds did not persist. There must have been many different types of molecules and many failures.

The Living Cell

The enclosure of self-copying compounds within a semi-permeable membrane that contains them but lets food pass through it, is the basis of the structure of a living cell, and the achievement of this stage of organization may be regarded as the origin of life.

The life of the cell consists in assimilation into itself of other materials, compensation for wastage by self-copying of its contained compounds, and reproduction by simple division. All these processes involve expenditure of energy and organisms require systems that obtain it.

The earliest organisms were dependent on the chemical substances in the environment for building themselves up, deriving the necessary energy from inorganic chemical reactions, probably without using oxygen as this gas was then rare in the atmosphere.

This stage may be illustrated by the bacteria. It is probable that originally there were many different ways of obtaining energy, as shown even today in the bacteria that oxidize iron or sulphur. The different proportions in which the isotopes of sulphur are found in deposits of different age give reason for believing that sulphur bacteria were active in the Precambrian Period, about 800 million years ago. (H. G. Thode, J. Macnamara, and H. W. Fleming.)

In bacteria the genetic material represented by deoxyribose nucleic acid molecules is distributed in the cell and not organized into a nucleus.

Plants and Animals

A higher stage of organization was reached when the genetical material was enclosed within a single nucleus inside each cell, and cell-division took place by means of a process called *mitosis* which ensures precise replication of the genetical material, now taking the form of chromosomes, in each daughter-cell.

Also at an early stage a new type of organization was evolved when, in addition to the absorption in water of salts containing nitrogen, phosphorus, etc., use was made of solar energy and substances like chlorophyll to fix carbon dioxide and to build it up by photosynthesis into carbohydrates (sugars and starches) which provide the main store of energy. Organisms that do this are *plants*. Plant-proteins, carbohydrates, and fats are therefore synthesized from materials not forming part of other living organisms. This mode of living is called *autotrophic*.

All these functions are carried out in the single cells which constitute the simplest plants, but in the more highly evolved plants there are special organs which carry out the different functions : roots, through which water and salts are absorbed from the soil, stems for their transmission, and leaves containing chlorophyll through which carbon dioxide is absorbed from the air and, in sunlight, built up into carbohydrates. The plant's food materials come to it, in the water, in the air, or the soil in which it lives, or in the case of a parasite from the host plant on which it lives ; and it does not need to go in search of its food materials.

The simplest plants, like *Chrysomonas*, are single, nucleated cells with thin walls. In typical plants whether single cells like *Chlamydomonas* or multicellular like any tree, the cell-walls are thick and made of cellulose.

Some plants are *parasitic* and rely on their hosts for part or all of their food requirements. Mistletoe does its own photosynthesis but depends on its host for water and salts ; dodder, a complete parasite, takes the products of its host's photosynthesis as well.

From the earliest plants, other organisms evolved the habit of obtaining their building materials and supply of energy ready-made by feeding on the proteins, fats, and energy-rich carbohydrates provided by already-existing organisms, and lost the power of synthesizing their food from inorganic materials. This mode of living is called *heterotrophic* and is represented by Fungi and by Animals.

Fungi (moulds and mushrooms) have usually been regarded as plants of a special kind lacking chlorophyll. They are therefore dependent on a supply of organic carbon compounds for their food which they obtain either by parasitizing other living plants or animals, or from decomposition of dead organisms, a mode of nutrition known as the *saprophytic*. Fungi resemble plants in that they do not move in search of their food, and that they reproduce by spores ; but it is possible that they may represent a separate group of organisms on their own.

Animals are dependent for their food on other plants or animals. Animals therefore cannot have come into existence before plants were evolved ; and ultimately the entire Animal Kingdom is dependent for its life on the continued existence of the Plant Kingdom.

The typical animal goes in search of its food which it ingests actively. In order to get it, the animal has organs of perception, movement, and food-capture. In the simplest animals consisting of a single cell, the ingestion of food takes place on the surface as in *Amoeba*. In the higher animals the centre of the body is hollow and the food is collected inside. In all animals the food is digested (i.e. broken down into simpler chemical substances) before being built up again into the compounds of the kind specific for the particular animal.

Above a certain level of organization animals are bilaterally symmetrical, an adaptation to movement ' one end first ', this being the end where the nervous system is most highly developed and takes the form of a brain, to which information is supplied by special sense-organs serving the functions of taste, sight, touch, balance, hearing, and smell.

Degenerate animals have adopted sessile habits, waiting for food materials to be wafted to them, as in sponges and ascidians. Such animals have (in the adult) lost their bilateral symmetry, power of locomotion, and special sense-organs. Still more degenerate animals, such as tape-worms, live as parasites inside other animals from which they derive all their nourishment. Some parasites (e.g. *Sacculina*) have lost all their special structures and are simply absorbent sacs producing germ-cells.

Viruses

Viruses are a category of very minute organisms, many times smaller than bacteria, which can multiply only within the living cells of susceptible hosts which they parasitize. Examples are the viruses of smallpox, measles, influenza, or poliomyelitis in man, distemper in dogs, foot-and-mouth disease in cattle, psittacosis in birds, mosaic disease of tobacco, and phages of bacteria.

Viruses consist of little more than particles of protein combined with deoxyribose nucleic acid which, like genes which they probably are, induce the formation of molecules similar to themselves out of the materials of the host cells. In the case of some plant viruses it has been shown that the organism consists of a pure molecule of protein. Viruses vary by mutation (which is why the nature of influenza changes), and are acted upon by natural selection resulting in adaptation. Viruses which kill their hosts are imperfectly adapted ; those which do not are well adapted.

Because viruses cannot reproduce themselves from anything but living matter it is difficult to regard them as survivors from the pre-cellular stage of evolution. More probably they must be regarded either as parasites evolved by degeneration from organisms like bacteria capable of synthesizing their substance from inorganic matter, or as descendants of what were originally parts of the host cell which have acquired separate existence.

There is no information about the time when viruses first appeared, but monkeys and men must have ‘ inherited ’ their herpes viruses from their common ancestor, 40 million years ago. (F. M. Burnet.)

The Evolution of Sex

In the simplest organisms consisting of a single cell, fusion between two individuals can take place, especially when environmental conditions become adverse. This process, which is a form of mutual ingestion, increases vigour and confers survival value. The two cells fuse together, and their two nuclei combine into one. By this means the hereditary units in the nuclei, the genes of the two original individuals, are pooled. This process known as *syngamy* is the basis of sex.

Subsequently the fused individuals perform reproduction by simple fission of their bodies into two. Their descendants will eventually fuse with the bodies of others and then reproduce by fission again. The repeated alternate syngamy and reproduction by fission is the basis of segregation and recombination of genes; genes are exchanged and recombined throughout the population; this process made possible the very much more rapid evolution of all higher plants and animals because the amount of possible variation due to previous mutations of the genes was thereby enormously increased.

At the early stages of the evolution of sex there were no separate sexes. The origin of difference between the sexes was the result of a variation whereby one type of fusing cell or *gamete* retained its power of locomotion by means of a flagellum and became a *sperm*, while the other type accumulated food-products, usually yolk, in its body which grew larger in consequence and lost its power of locomotion, becoming the *egg*. This division of function, in which the locomotion of the sperm enabling it to swim to and fuse with (fertilize) the egg was combined with the capacity of the egg to provide a store of food, conferred survival value on the new fused individual (fertilized egg or *zygote*).

All subsequent evolution of differences between the sexes in plants and animals is based on this fundamental differentiation between sperm-producers or *males* and egg-producers or *females*.

It involves the differentiation between the organs that produce sperm (antheridia in plants, testes in animals) and those that produce eggs (archegonia in plants; ovaries in animals). In flowering plants where the sperms and eggs are incorporated within the spores, it involves the differentiation between floral leaves that produce pollen (stamens) and ovules containing the embryo-sacs (carpels). It involves the differentiation between the ducts in animals that evacuate the sperms (vasa deferentia) and the eggs (oviducts). It is also responsible for the difference in behaviour between males and females, leading to the actions which result in the fertilization of eggs by sperms.

West side: fifth bay
Right-hand case

Multicellular Organisms, Embryonic Development, and Death

The evolution of many-celled organisms, by internal compartmentalization of single-celled organisms, was a step of major importance because it allowed organisms to grow to larger size and made possible the differentiation of different parts of the body, specialized for the performance of different functions which previously had all to be carried out by the single cell. This differentiation increases the efficiency of the performance. The evolution of multicellular status occurred independently in plants and animals, and probably also in fungi and in sponges.

In plants the leaves specialize as light and carbon dioxide absorbers, the roots as water and salt absorbers, and the stems as supporting and conducting structures. In animals the outermost layer serves for protection and supplies the tissues for nervous conduction and sense perception, the innermost layer for digestion and absorption, and other tissues carry out the functions of respiration, excretion, locomotion, skeletal support, and reproduction.

A consequence of the evolution of the many-celled organism is that the fertilized egg, being a single cell, has to undergo a process called *embryonic development* in order to return to the complexity and size of its parents which can now be called *adults*. It divides into more and more cells, and they become *differentiated* into the different regions of the body, while *growth* enables the embryo to regain the size of its parents. The provision of reserve food-materials in the egg is of increasing importance in this process until in the highest group of all, the viviparous mammals, the resources of the mother are substituted for the yolk in the egg.

The subdivision of the body into many cells imposed another consequence of major importance : only the reproductive cells, being single cells, can fuse with each other and pool their genes at the start of the next generation : all the remainder of the body of the individual of the previous generation gradually breaks down and dies sooner or later. This is to be contrasted with the condition in the lowest organisms where the body consists of a single cell. Their reproduction is always by division and the old individual becomes directly transformed into the new individuals. Since in principle this process can be repeated indefinitely, such organisms are said to be potentially immortal.

Multicellular plants can multiply by spores or offsets, and, where the complexity of the body is not too great, multicellular animals can reproduce by simple division, as in sea-anemones and worms. In some cases eggs can develop without fertilization, by *parthenogenesis* ; but the products of all such types of reproduction are deprived of the potential advantages of genetic recombination.

The Colonization of Land

Life originated in water, and the properties of this medium imposed certain restrictions on the evolutionary possibilities open to plants which never advanced further in water than the stage represented by sea-weeds. Animals were able to rise to higher levels of organization represented by the octopus and the fish. The colonization of land in the Devonian Period provided the opportunities for the highest development in both the Plant and Animal Kingdoms.

The conquest of the land was started by organisms living in shallow water subject to periodical desiccation. In both plants and animals there were two stages before their adaptation to life on land could be completed. The first adaptation enabled the organisms to live in air, and this involved ability to breathe atmospheric oxygen, to resist desiccation, and to obtain food-materials. Plants achieved it by the development of a resistant outer layer perforated for breathing, supporting structures (stems), and organs for absorbing food from the ground (roots).

In animals (snails, insects, vertebrates) it was achieved similarly by the development of a resistant skin and of skeletal structures, and by the formation of cavities from which oxygen could be absorbed into the body (mantle-cavities, tracheae, and lungs respectively). The animals' needs of locomotion were met by the modification of appendages : the 'foot' of snails, legs of insects, and five-toed limbs of vertebrates.

The second adaptation was concerned with reproduction. In order to reach the egg and fertilize it, the sperm must swim through a film of water. For this reason mosses, ferns, and horsetails cannot escape from damp areas where there is a sufficient film of moisture. Among animals, Amphibia require actual ponds or rivers in which the eggs and sperms are spawned and the earliest stages of development are passed.

In plants the difficulty was met by developments in connexion with the phenomenon of alteration of generations, as a result of which the eggs and sperms are incorporated inside the pistil and the anthers of the flowers, and pollination takes place through the air. This enabled flowering plants to colonize *dry* land all over the earth.

In animals the problem of adaptation to dry land was solved by the practice of internal fertilization inside the body of the female, and the production of embryonic membranes containing fluids, and a shell surrounding the fertilized egg. A further refinement found in the higher mammals and a few other forms is the retention of the fertilized egg and its membranes inside the body of the mother where development takes place, thereby dispensing with the shell.

Alternation of generations in plants

Reproduction may follow one or other of two methods. In the first, known as *asexual*, a single cell or spore develops without fertilization into a plant called a *gametophyte*. This on the other hand follows the second or *sexual* method and produces sperms and eggs; and the egg, after fertilization, develops into a plant called a *sporophyte*. This produces spores and the cycle of an asexual generation alternating with a sexual generation is thereby completed. In some algae these two generations look alike.

In mosses the visible 'plant' is the gametophyte generation which produces sperms and eggs. These after fertilization in a film of moisture produce the inconspicuous sporophyte generation which remains attached to the parent and produces the spores from which a new gametophyte generation develops without fertilization.

In ferns the visible 'plant' is the sporophyte generation which produces spores. These develop without fertilization in damp surroundings into the very small independent gametophyte generation (called the *prothallus*) which produces sperms and eggs. These after fertilization give rise to the sporophyte generation.

In flowering plants the visible 'plant' is the sporophyte generation, and the gametophyte generation is represented only by a few nuclei, telescoped into a stage in the formation of the spores. The sperm is the nucleus of the pollen-grain and the egg is enclosed in the ovule within the pistil. The sexual differentiation of sperm and egg has become impressed on the sporophyte generation, with the result that there are microspores (pollen-grains) and macrospores (embryo-sacs), produced respectively by stamens and carpels which are the 'male' and 'female' constituents of flowers. Pollination leads to the development of the fertilized egg into a seed which is a 'packaged' dormant embryo equipped for wide dispersal.

The progressively greater dominance of the sporophyte over the gametophyte generation in evolution is a consequence of the fact that the sporophyte, produced by the fertilization of gametes, contains the diploid number of chromosomes. The number is reduced (by *meiotic division*) to half at the formation of spores with the result that the gametophyte contains only the haploid number. The sporophyte therefore contains twice as many genes and is capable of more variation than the gametophyte.

The Transition from Water to Land

The majority of organisms that live in water are so different from those that live on land, that it is not always realized that the transition from one medium to the other was easy and involved little in the way of new structures and no discontinuity in the essential functions of life. This may conveniently be illustrated by the evolution of fishes into amphibia.

In fishes respiration is carried out by taking water into the mouth, closing it, lifting its floor and forcing the water out through the gill-slits where the blood-vessels absorb the oxygen in solution. Locomotion is carried out by rhythmical contractions of the muscles of the side of the body giving rise to undulatory movements which press the surrounding water backwards, while the fins serve for deflection.

In some fishes, including living forms, the hindmost pair of gill-slits are modified into sacs which become lungs and serve to absorb oxygen from air when the water or mud in which the fish lives dries up. This is the situation in amphibia where air is forced into the lungs in exactly the same way as water was forced through the gill-slits in the fish. The same muscles and nerves were used during and after the transition without any functional discontinuity. The same is true for locomotion, because the undulatory movements of the fish are the same as those which cause the limbs to move backwards and forwards in an amphibian like the newt, for the limbs stick out at right angles to the body and serve as oars to row the animal along while dragging its belly on the ground. The same neuro-muscular mechanism and pattern of behaviour survived the transition without any functional discontinuity. The only novelty was the reduction of the large number of radials of the fin of the fish to five and their separation into fingers and toes, characteristic of all land animals from amphibia to man inclusive.

The smoothness with which this and other transitions was accomplished was made possible by the fact that in possessing certain structures capable of changed functions, the fishes whose descendants made the transition happened to be ' pre-adapted ' to the new medium in which they found themselves (see p. 22). This example shows how evolution proceeded as a gradual change by a series of improvizations which, if serviceable, were favoured, preserved, and improved upon by natural selection.

Psycho-social evolution

Up to the evolution of man from earlier forms, the mechanism of evolution has been natural selection of fortuitous variations which were the result of mutation and recombinations of different genes.

Among some birds and mammals there are beginnings of tradition taught by one generation to the next. In them, simple situations are mastered through behaviour taught by an older or more experienced member of their own species. Chaffinches learn part of their song by copying adult chaffinches. Geese teach their goslings where danger lurks from foxes. Lions teach their cubs methods of hunting. Chimpanzees show their young many simple things by example and encouragement. This transmission of experience by education is a process additional to natural selection but, like it, can confer survival value.

The evolutionary lineage leading to man was characterized by specialization of the central nervous system, producing increased intelligence and conferring survival value with each improvement. With increasing delay in development, prolongation of childhood, and consolidation of the family as a biological unit, man became increasingly gregarious and social. His awareness, ability to think, to express his thoughts in speech, and to understand, enabled him to exchange experience, store it in memory and tradition, and impart it to the young by a cumulative practice of culture and heredity called psycho-social evolution. The result was that, when confronted with problems, man became increasingly able to choose a line of conduct by personal and collective decision with conscious purpose.

The importance of psycho-social factors in the evolution of Man has been outlined on page 21. (Sir Julian Huxley.)

West side: fifth bay
Facing wall

On the facing wall of this Bay are 6 panels, each illustrative of one of the subjects treated in this Bay. The panel relating to the evolution of sex is based on a microphotograph made by Mr. J. Smiles of the National Institute for Medical Research of a fertilized mouse egg, kindly lent by him and Dr. A. S. Parkes, C.B.E., F.R.S.

Literature for Further Reading

General Works

BALDWIN, E. 1948. *An introduction to comparative biochemistry*. Cambridge.

BARLOW, NORA (editor) 1934. *Charles Darwin's diary of the voyage of H.M.S. Beagle*. Cambridge.

—— (editor) 1945. *Charles Darwin and the voyage of the Beagle*. London.

—— (editor) 1963. Darwin's ornithological notes. *Bull. Br. Mus. nat. Hist*. Historical series **2** 203–278.

BATES, M. and HUMPHREY, P. M. (editors) 1957. *The Darwin reader*. London.

BLUM, H. F. 1955. *Time's arrow and evolution*. Princeton.

BONNER, J. T. 1958. *The evolution of development*. Cambridge.

CAIN, A. J. 1954. *Animal species and their evolution*. London.

CARTER, G. S. 1951. *Animal evolution*. London.

—— 1957. *A hundred years of evolution*. London.

CLARK, F. and SYNGE, R. L. M. (editors: English-French-German edn.) 1959. *Proceedings of the first international symposium on the origin of life on the earth, held at Moscow* 19–24 *August*, 1957. London.

CLARK, J. G. D. 1952. *Prehistoric Europe; the economic basis*. London.

COLE, SONIA 1959. *The Neolithic revolution*. London.

—— 1963. *The races of man*. London.

CUÉNOT, L. 1925. *L'adaptation*. Paris.

DARLINGTON, C. D. 1953. *The facts of life*. London.

—— 1958. *Evolution of genetical systems*. Edinburgh.

DARLINGTON, P. J. 1957. *Zoogeography*. New York.

DARWIN, C. (reprinted) 1952. *Journal of researches into the geology and natural history of the countries visited during the voyage of H.M.S. Beagle*, London, 1839. New York—London.

—— (reprinted) 1956. *Journal of researches into the natural history and geology of the countries visited during the voyage of H.M.S. Beagle*, London, 1845. With an introduction by Sir Gavin de Beer. Cambridge.

—— (6th edn. 1872 reprinted: The World's Classics) 1956. *The origin of species*. With an introduction by Sir Gavin de Beer. Oxford.

—— (reprinted) 1958. *Sketch* written in 1842, and *Essay* written in 1844 in *Evolution by natural selection*. With a foreword by Sir Gavin de Beer. Cambridge.

DARWIN C. and WALLACE, A. R. (reprinted) 1958. *On the tendency of species to form varieties; and on the perpetuation of varieties and species by natural means of selection, London* 1870, in *Evolution by natural selection*. With a foreword by Sir Gavin de Beer. Cambridge.

DE BEER, GAVIN R. (editor) 1938. *Evolution*. Oxford.
—— (3rd edn.) 1958. *Embryos and ancestors*. Oxford.
—— (editor) 1959. Darwin's journal. *Bull. Br. Mus. nat. Hist.* Historical series **2** 1—21.
—— (editor) 1960. Notebooks on transmutation of species. I–IV. *Bull. Br. Mus. nat. Hist.* Historical series **2** 23–183
—— 1963. *Charles Darwin*. London.
—— 1964. *Atlas of evolution*. London.
DOBZHANSKY, T. 1951. *Genetics and the origin of species*. 3rd ed. New York.
—— 1962. *Mankind evolving*. New Haven.
DOWDESWELL, W. H. 1955 *The mechanism of evolution*. London.
FISHER, R. A. 1930. *The genetical theory of natural selection*. Oxford.
FORD, E. B. 1957. *Mendelism and evolution*. London.
—— 1964. *Evological genetics*. London.
HALDANE, J. B. S. 1932. *The causes of evolution*. London.
HUXLEY, J. S. (editor) 1940. *The new systematics*. Oxford.
—— 1942. *Evolution: the modern synthesis*. London.
—— 1953. *Evolution in action*. London.
—— (editor) 1954. *Evolution as a process*. London.
—— and T. H. 1947. *Evolution and ethics*, 1893–1943. London.
JOHNSON, M. L., ABERCROMBIE, M. and FOGG, G. E. (editors) 1954. The origin of life. *New Biol.* No. 16. London.
LAMARCK, J. B. DE (reprinted) 1873. *Philosophie zoologique*. Paris.
LE GROS CLARK, WILFRED 1962. *History of the primates*. 8th ed. London.
MAYR, E. 1942. *Systematics and the origin of species*. New York.
—— (editor) 1957. *The species problem*. Washington.
—— 1963. *Animal species and evolution*. Oxford.
OAKLEY, K. P. 1963. *Man the tool-maker*. 5th ed. London.
—— and MUIR-WOOD, H. M. 1956. *The succession of life through geological time*. London.
PRINGLE, J. W. S. and others 1953. Evolution. *Symp. Soc. exp. Biol.* No. 7 1–448.
RENSCH, B. 1954. *Neuere Probleme der Abstammungslehre*. 2nd ed. Stuttgart.
—— 1959. *Evolution above the species level*. London.
SHEPPARD, P. M. 1958. *Natural selection and heredity*. London.
SIMPSON, G. G. 1950. *The meaning of evolution*. London.
—— 1953. *The major features of evolution*. New York.
—— 1964. *This view of life. The world of an evolutionist*. New York.
STEBBINGS, J. L. 1950. *Variation and evolution in plants*, New York.
TINBERGEN, N. 1951. *The study of instinct*. Oxford.
WADDINGTON, C. H. 1939. *An introduction to modern genetics*. London.
—— 1957. *The strategy of the genes*. London.
WALLACE. A. R. 1870. *Contributions to the theory of natural selection*. London.

Specialist Works

Allison, A. C. 1956. Human haemoglobin types. *New Biol.* No. 21 43–58.

Barghoorn, E. S. 1957. Origin of Life. *Mem. geol. Soc. Am.* No. 67 **2** 75–85.

Boyd, W. C. 1950. *Genetics and the races of man.* Boston.

Burnet, F. M. 1958. *Viruses and man.* Harmondsworth.

Cain, A. J. and Sheppard, P. M. 1952. The effects of natural selection on body colour in the land snail *Cepaea nemoralis. Heredity* **6** 217–231.

Clausen, J., Keck, D. D., and Hiesey, W. M. 1940–1958. Experimental studies on the nature of species. *Publs Carnegie Instn.* Nos. 520, 564, 581, and 615; 1—452, 1–174, 1–129, 1–312.

Darlington, C. D. 1947. The genetic components of language. *Heredity* **1** 269–286.

Darwin, F. (editor) 1887. *The life and letters of Charles Darwin, including an autobiographical chapter*. London.

de Beer, Gavin 1954. *Archaeopteryx* and evolution. *Advmt. Sci.* **42** 160–170.

Diver, C. R. 1929. Fossil records of mendelian mutants. *Nature, Lond.* **124** 183.

Fisher, J. and Lockley, R. M. 1954. *Sea birds*. London.

Fisher, R. A. 1931. *The evolution of dominance. Biol. Rev.* **6** 345–368.

—— 1932. The bearing of genetics on theories of evolution. *Sci. Progr.* **27** 273–287.

—— 1950. The ' Sewall Wright Effect '. *Heredity* **4** 117–119.

Ford, E. B. 1936. The genetics of *Papilio dardanus. Trans. R. ent. Soc., Lond.* **85** 435–466.

—— 1940. Genetic research in the Lepidoptera. *Ann. Eugen.* **10** 227–253.

—— 1952. The influence of radiation on the human genotype, in *Biological Hazards of Atomic Energy* edited by A. Haddow. Oxford.

—— 1956. *Genetics for the medical student*. London.

Harland, S. C. 1936. The genetical conception of species. *Biol. Rev.* **11** 83–112.

Hélbaek, H. 1959. Domestication of food plants in the Old World. *Science* n. s. **130** 365–372.

Hovanitz, W. 1948. Differences in the field activity of two female color phases of *Colias* butterflies at various times of the day. *Contr. Lab. verteb. Biol. Univ. Mich.* 1–37.

Huxley, T. H. 1878. Evolution in biology. *Encyclopaedia Britannica*, 9th edn. **8** 744–751.

Huxley, J. S. 1914. The courtship habits of the Great Crested Grebe. *Proc. Zool. Soc., Lond.* 491–562.

Kettlewell, H. B. D. 1956. A résumé of investigations on the evolution of melanism in the Lepidoptera. *Proc. R. Soc.* **145** B 297–306.

LACK, D., 1947. *Darwin's finches*. Cambridge.

LANKASTER, E. R. 1912. *Science from an easy chair*. 2nd ser. London.

LISSMANN, H. W. 1951. Continuous electrical signals from the tail of a fish. *Nature, Lond.* **167** 201–202.

MCMEEKAN, C. P. 1940. Growth and development in the pig, with special reference to carcass quality characters. *J. agric. Sci. Camb.* **30** 276–343.

MARSDEN-JONES, E. M. and TURRILL, W. B. 1938. Fifth report of the transplant experiments. *J. Ecol.* **26** 359–379.

MEINERTZHAGEN, R. 1954. *Birds of Arabia*. Edinburgh.

MILLER, S. L. 1953. A production of amino-acids under possible primitive earth conditions. *Science* **117** 528–529.

MORGAN, T. H., BRIDGES, C. B. and STURTEVANT, A. H. 1925. The genetics of *Drosophila*. *Bibliogr. genet.* **2** 1–262.

MOURANT, A. E. 1950 *The distribution of the human blood groups*. Oxford.

MULLER, H. J. 1947. The gene. *Proc. R. Soc.* **134** B 1–37.

MUNTZING, A. 1930. Outline to genetic monograph of the genus *Galeopsis* with special reference to the nature of inheritance of partial sterility. *Hereditas* **13** 185–341.

PARKER, W. H. 1956. Species transgression in one horizon. *Publ. Syst. Ass.* No. **2** 9–15.

SACHS, L. 1952. Polyploid evolution and mammalian chromosomes. *Heredity* **6** 257–364.

SIMPSON, G. G. 1951. *Horses*. Oxford.

SPATH, L. F. 1938. *A Catalogue of the ammonites of the Liassic family Liparoceratidae in the British Museum (Natural History)*. London.

SPOONER, E. T. C. and STOCKER, B. A. D. (editors) 1956. *Bacterial anatomy*. Cambridge.

STEBBINS, R. C. 1949. Specialisation in salamanders of the Plethodontid genus *Ensatina*. *Univ. Calif. Publ. Zool.* **48** 2377–526.

THODE, H. G., MACNAMARA, J. and FLEMING, H. W. 1953. Sulphur isotope fractionation in nature and geological and biological time scales. *Geochim. cosmochim. Acta* **3** 235–243.

THORPE, W. H. 1958. The learning of song patterns by birds with especial reference to the song of the chaffinch *Fringilla coelebs*. *Ibis* **100** 535–570.

TYLER, S. A. and BARGHOORN, E. S. 1954. Occurrence of structurally preserved plants in Pre-Cambrian rocks. *Science* **119** 606–608.

WADDINGTON, C. H. 1953. The evolution of adaptations. *Endeavour* **12** 134–139.

WATSON, J. D. and CRICK, F. H. C. 1953. Molecular structure of nucleic acids. *Nature, Lond.* **171** 737–738.

WILLMER, E. N. 1949. Some evolutionary aspects of mammalian colour-vision. *Proc. Linn. Soc., Lond.* **161** 97–111.

GEOLOGICAL TIME-SCALE†

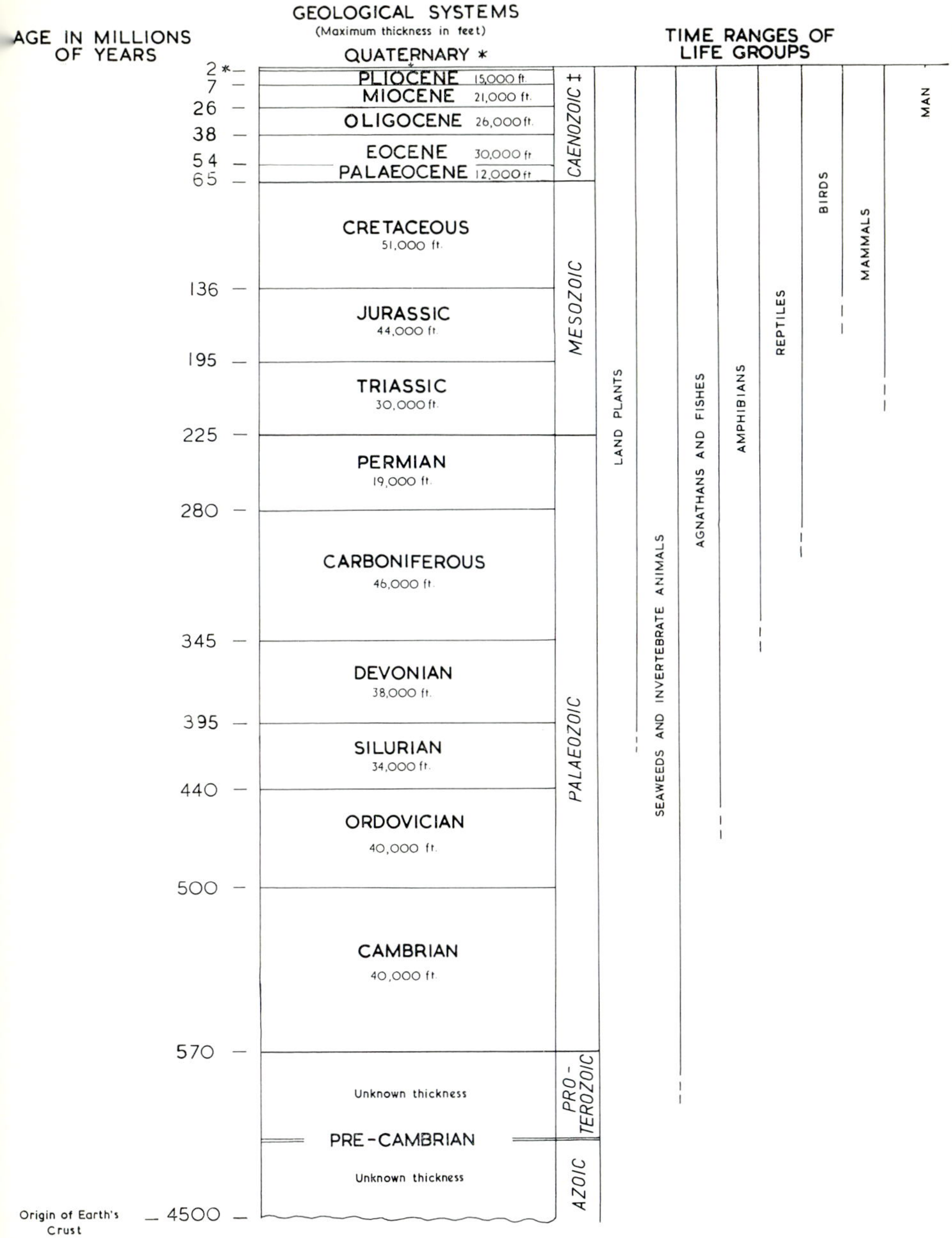

† Time-scale approximate with probable error of ± 5% throughout
* Quaternary (Pleistocene and Holocene) 6,000 feet +
‡ Caenozoic = Tertiary (Palaeocene - Pliocene) + Quaternary
Column proportional to time-scale

Index

BRITISH MUSEUM (NATURAL HISTORY)

DEPARTMENT OF PALAEONTOLOGY

HANDBOOKS

The Succession of Life through Geological Time. By K. P. Oakley & H. M. Muir-Wood; 7th edition; pp. vii, 96; coloured frontispiece; 12 plates, 4 maps, 16 text-figs. 1967. 8vo. 5*s.* (postage and packing 1*s.* 6*d.* overseas 1*s.* 9*d.*)

Man the Toolmaker. By K. P. Oakley ; 5th edition (reprinted) ; pp. vi, 98 ; 2 plates, 41 text-figs. 1965. 8vo. 5*s.* (postage and packing 1*s.* overseas 1*s.* 3*d.*)

History of the Primates : An Introduction to the Study of Fossil Man. By W. E. Le Gros Clark ; 9th edition ; pp. viii, 127 ; 45 text-figs. 1965. 8vo. 7*s.* 6*d.* (postage and packing 1*s.* 3*d.* overseas 1*s.* 6*d.*)

The Neolithic Revolution. By Sonia Cole ; 4th edition ; pp. x, 66 ; 18 plates, 33 text-figs. 1967. 8vo. 5*s.* (postage and packing 1*s.* overseas 1*s.* 3*d.*)

Digging up Bones. By D. R. Brothwell. pp. xiii, 194 ; 18 plates, 64 text-figs. 1963. (Reprinted 1965.) 19*s.* 6*d.* (postage and packing 2*s.* overseas 2*s.* 6*d.*)

Races of Man. By Sonia Cole. 2nd edition ; pp. 131 ; 12 plates, 34 text-figs. 1965. 11*s.* 6*d.* (postage and packing 1*s.* 9*d.* overseas 2*s.*)

Fossil Amphibians and Reptiles. By W. E. Swinton ; 4th edition ; pp. ix, 133 ; 17 plates (1 coloured) 68 text-figs. 1965. 8vo. 5*s.* (postage and packing 1*s.* 6*d.* overseas 1*s.* 9*d.*)

Fossil Birds. By W. E. Swinton ; 2nd edition ; pp. vi, 63 ; coloured frontispiece ; 11 plates, 25 text-figs. 1965. 8vo. 7*s.* 6*d.* (postage and packing 1*s.* 3*d.* overseas 1*s.* 6*d.*)

Dinosaurs. 4th edition; By W. E. Swinton; pp. xiv, 44, frontispiece; 11 plates (2 coloured); 6 text-figs. 1969. 8vo. 5*s.* (postage and packing 1*s.* overseas 1*s.* 3*d.*)

British Mesozoic Fossils. 3rd edition ; pp. v, 207 ; 73 plates and coloured map, 1967. 8vo. 12*s.* 6*d.* (postage and packing 1*s.* 9*d.* overseas 2*s.*)

British Caenozoic Fossils. 3rd edition; pp. vii, 132; 44 plates. Coloured map, 1968. 8vo. 6*s.* (postage and packing 1*s.* 3*d.* overseas 1*s.* 6*d.*)

British Palaeozoic Fossils. 3rd edition ; pp. vii, 208 ; 69 plates. Coloured map. 1969. 8vo. 13*s.* (postage and packing 1*s.* 9*d.* overseas 2*s.*)

An Early History of Palaeontology. By W. N. Edwards ; pp. viii, 58 ; 4 plates, 19 text-figs. 1967 8vo. 5*s.* (postage and packing 1*s.* 3*d.* overseas, 1*s.* 6*d.*)

Postage rates shown are correct at the time of going to press.